Collins

Science

KS3

Science

Year 8
workbook

KS3

Science

Year 8
workbook

Francesca Walsh, Dan Evans & Ron Holt

Contents

Biology

Chemistry

Physics

Answers

Acknowledgements

Cover & p1 © Irina1977/Shutterstock.com, © 1stGallery/Shutterstock.com; p1 © Riyazi/Shutterstock.com;

p2 © R. Gino Santa Maria/Shutterstock.com; p9 © aleonello calvetti/Shutterstock.com;

p10 © Shutterstock.com/BioMedical; p11 © bikeriderlondon/Shutterstock.com;

p12 © phil Holmes/Shutterstock.com; p14 © Alila Medical Media/Shutterstock.com;

p15 © Mikhail Starodubov/Shutterstock.com; p36 © Chaikovskiy Igor/Shutterstock.com;

P36© andreiuc88/Shutterstock.com; p38 © Picsfive/Shutterstock.com; p40 © Picsfive/Shutterstock.com;

p46 © Zern Liew/Shutterstock.com; p58 © Elenarts/Shutterstock.com

Vocabulary Builder

1 This question is about respiration.
Draw lines to match each word to its correct definition.

Word	Definition
Aerobic respiration	Process that converts sugar into alcohol
Mitochondria	Where respiration takes place
Anaerobic respiration	Respiration using oxygen
Lactic acid	Respiration in the absence of oxygen
Fermentation	Substance that builds up in muscles during anaerobic respiration

[4]

2 For each of the statements below, write the term that is being defined.

a) An animal with a backbone _____ [1]

b) The soft tissue found inside bones _____ [1]

c) The protein that ligaments are made from _____ [1]

d) The type of muscle found in the heart _____ [1]

e) The organelle that is the site of aerobic respiration _____ [1]

f) The form in which glucose is stored in muscles and the liver _____ [1]

g) Drugs that athletes take to increase muscle strength _____ [1]

3 The following passage is about the skeleton.
Fill in the spaces using words from the box.

skull	cartilage	blood cells	move	ligaments
heart	support	synovial fluid	lungs	joints

The skeleton has four functions. The first is protection; for example, the rib cage protects the _____ and the _____, whilst the _____ protects the brain. The second function is to provide _____. The third function is that it allows us to _____. The final function is that new _____ are made in the bone marrow.

Movement occurs at _____ in the body. The ends of our bones are covered with a tough, rubbery substance called _____, which protects them. _____ cushions and lubricates joints. The bones are held together by tough straps called _____ .   [10]

4 Muscles work in pairs against each other.

a) What do we call pairs of muscles like this? _____ [1]

b) In each of the lists below, there is a pair of muscles and an **odd one out**.
   Circle the odd one out in each list.

   i) quadriceps   triceps   biceps   [1]

   ii) hamstring   quadriceps   deltoid   [1]

5 Write each part of the body from the box below in the correct column in the table to show the type of joint present.

elbow   shoulder   knee   neck   finger   wrist   hip

Hinge Joints	Ball and Socket Joints	Gliding / Pivot Joints

[7]

6 In each of the following lists, circle the **odd one out**. Give a reason for your answer.

a) humerus   sternum   radius   ulna

   Reason: _____ [2]

b) femur   tibia   fibula   scapula

   Reason: _____ [2]

Total Marks _____ / 35

1 The table below shows the composition of inhaled and exhaled air.

Gas	% in Inhaled Air	% in Exhaled Air
Oxygen	21	16
Carbon dioxide	0.04	5
Nitrogen	79	79

a) Draw a pie chart in the circle below to show the composition of **exhaled** air.

The Composition of Exhaled Air

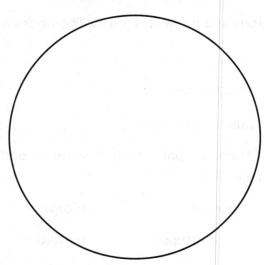

[3]

b) Which of the statements below is **true**? Underline the correct one.

    A: About half of the oxygen inhaled is used in respiration.

    B: About a quarter of the oxygen inhaled is used in respiration.

    C: About a tenth of the oxygen inhaled is used in respiration. [1]

2 Two groups of students wanted to find out how temperature affected yeast fermentation. They took five conical flasks and into each measured 5g glucose, 50cm^3 of warm water and 1g of yeast. They put a bung in the top, put each flask at a different temperature and left them for a week.

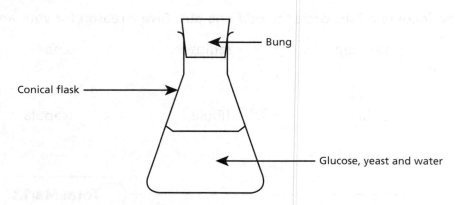

Bung

Conical flask

Glucose, yeast and water

After a week, they measured the specific gravity of the mixture in the conical flask.
The higher the specific gravity, the more alcohol is present.
Their results are shown in the table below.

Group	Specific Gravity of Mixture at Different Temperatures (arbitrary units)					
	20°C	25°C	30°C	35°C	40°C	45°C
1	1.5	2.5	4.0	4.2	3.5	1.0
2	2.0	1.0	3.5	4.0	2.0	1.0

a) On the grid below, plot a line graph of Group 1's results. [6]

b) Label both axes. [2]

c) Join your points. [1]

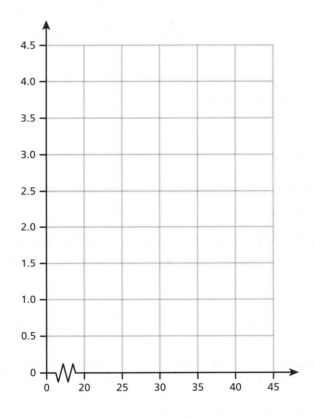

d) Use your graph to estimate the specific gravity of a mixture
incubated at 27.5°C. .... [1]

e) One of Group 2's results is anomalous.
Write the temperature of the anomalous result. .... [1]

f) At what temperature was most alcohol produced? .... [1]

3 An athlete was asked to run a distance of 10 000 metres. Every 2000 metres, she increased her speed. The graph below shows the levels of lactic acid in her leg muscles during the exercise period. The table gives information on the athlete's speed (in metres per second).

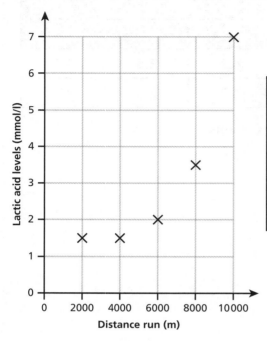

Distance Run (m)	Speed (m/s)
0 < Distance ≤ 2000	3.5
2000 < Distance ≤ 4000	3.8
4000 < Distance ≤ 6000	4.1
6000 < Distance ≤ 8000	4.4
8000 < Distance ≤ 10 000	4.9

a) Join the points on the graph. [1]

b) Use the graph to estimate the athlete's lactic acid levels after she had run 9000 metres. .... [1]

c) What is the maximum speed at which the athlete can run before lactic acid begins to accumulate in her leg muscles faster than it can be removed? .... [1]

Total Marks .... / 19

Testing Understanding

1 The following statements are about respiration.
For each statement, indicate whether it refers to **aerobic** or **anaerobic** respiration.

a) Uses oxygen .... [1]

b) Produces lactic acid .... [1]

c) Produces the most energy .... [1]

d) Produces carbon dioxide and water .... [1]

e) Takes place in the cytoplasm .... [1]

2 Anaerobic respiration happens in some microbes and also in humans.
Write the word equation for anaerobic respiration in humans.

_____ [3]

3 a) Fill in the missing gaps to complete the word equation for aerobic respiration.

Glucose + _____ $\longrightarrow$ _____ + Water + Energy   [2]

b) How are the substances needed for respiration transported to the respiring cells?

_____ [1]

c) Name the cell structure in which aerobic respiration occurs.

_____ [1]

4 a) Yeast can respire aerobically and anaerobically.
What name is given to **anaerobic respiration** in yeast?

_____ [1]

b) Write the equation for this reaction using words from the box below.

ethanol	glucose	energy	carbon dioxide

_____ [4]

5 The picture below shows the bones and two muscles in the arm.

a) What is the name of muscle **A**?

_____ [1]

b) What is the name of muscle **B**?

_____ [1]

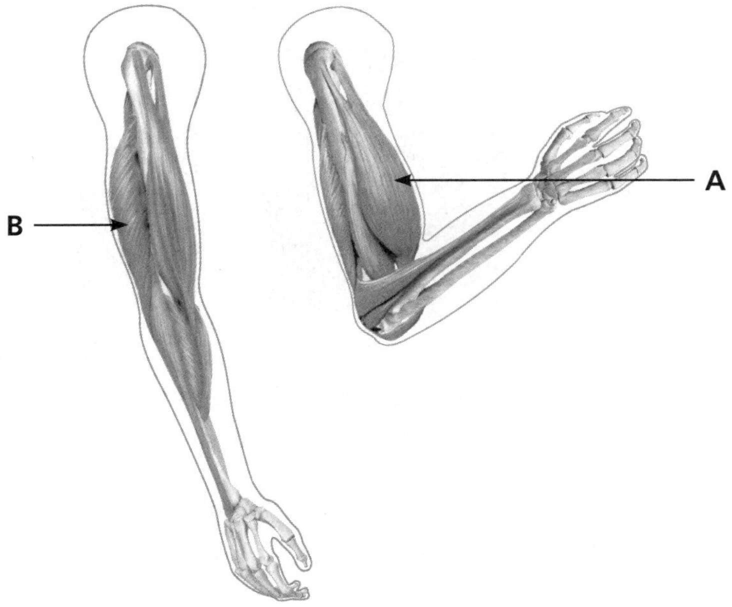

B ———→

A

c) What happens to each muscle in order to raise the arm?

....

.... [2]

d) What is the name of the structures that attach the muscle to the bone?

.... [1]

e) Circle the statements below that are likely to increase muscle strength.

torn muscle   **taking anabolic steroids**   **taking aspirin**

regular exercise   **a low carbohydrate diet**   [2]

6 Choose two words from the box to complete the sentence below.

| shorter | longer | fatter | thinner | rounded |

When muscles contract, they become .... and .... [2]

7 The picture below shows two joints in the human body.

a) On the diagram, write the name of each type of joint.

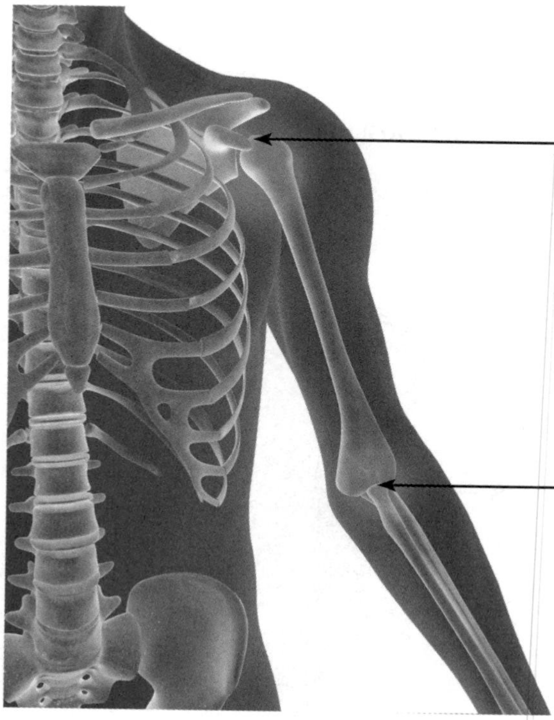

A: ....

B: ....

[2]

b) A third type of joint is a **pivot joint**.

Where in the body would you find a pivot joint?

.... [1]

8 As a result of aerobic respiration, the composition of the air that is breathed in and the air that is breathed out changes.

Underline the correct statement from the list below to show how the composition changes.

   A: The air that is breathed in contains more oxygen and more carbon dioxide than the air that is breathed out.

   B: The air that is breathed in contains more oxygen and less carbon dioxide than the air that is breathed out.

   C: The air that is breathed in contains less oxygen and more carbon dioxide than the air that is breathed out.

   D: The air that is breathed in contains less oxygen and less carbon dioxide than the air that is breathed out. [1]

9 **a)** When Anna dances, pairs of muscles in her arms and legs act **antagonistically**.
How do antagonistic muscles work? Underline the correct statement.

   A: Both muscles contract at the same time

   B: One muscle is big and one muscle is small

   C: One muscle is stronger and controls the weaker one

   D: When one muscle contracts, the other relaxes

   E: Both muscles relax at the same time [1]

b) When Kwame dances, his heart beats faster than when he is resting. Explain why.

_____

_____

_____

_____ [4]

10 The drawing below is of a human skeleton.

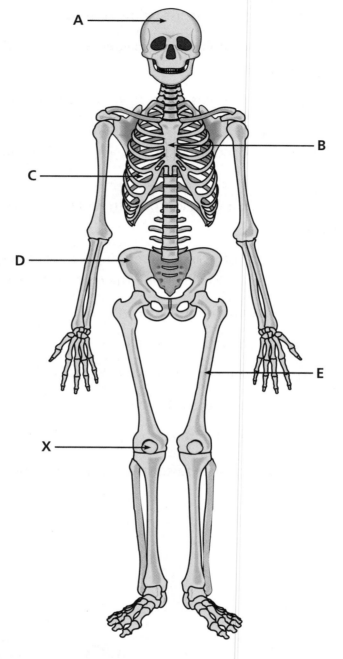

a) Name the parts labelled **A–E**.

A: ....   B: ....

C: ....   D: ....

E: ....   [5]

b) Name the type of joint shown by the letter **X**. ....   [1]

1 Zac wanted to prove that woodlice respire. He collected three boiling tubes and labelled them A, B and C. He placed some limewater in the bottom of each one. He then placed a piece of gauze halfway up each tube. Zac put five woodlice on the gauze in tube A and ten woodlice on the gauze in tube B. He did not place any woodlice in tube C. He sealed all the tubes with a bung and left them for one hour in the laboratory.

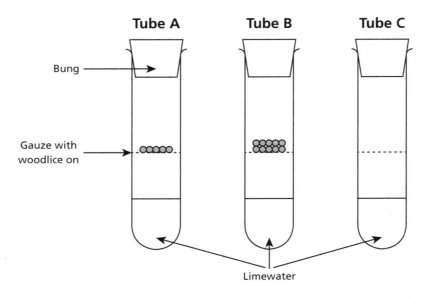

a) Why did Zac put limewater in the tubes?

.... [1]

b) In which tube would Zac first expect to see a change in the limewater?

.... [1]

c) What change would Zac expect to see in the limewater?

.... [1]

d) What name do we give to tube C?

.... [1]

e) Faisal suggested the experiment would work better if Zac placed the tubes in an incubator at 50°C.
Suggest why this is not a good idea.

....

.... [1]

Total Marks .... / 5

1 Read the passage below and then answer the questions that follow.

Osteoporosis is a condition that affects the bones, causing them to become weak and more likely to fracture (break). These fractures most commonly occur in the spine, wrist and hips, but can also affect other bones such as the arm or pelvis.

In childhood, bones grow and repair very quickly. They stop growing between the ages of 16 and 18 and, from about the age of 35, they begin to lose their density. For some people this can lead to osteoporosis.

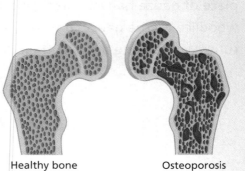

Healthy bone   Osteoporosis

Certain factors increase the risk of developing the condition, including long-term use of particular medications, heavy drinking and smoking, and failure of the body to absorb certain nutrients.

There is no cure for osteoporosis, but there are many treatments available. These include:

- drugs that slow down the rate at which the cells that break down bone work
- drugs that stimulate cells that create new bone
- calcium and vitamin D supplements.

a) Suggest why osteoporosis is rare in people under 40 years of age.

_____

_____ [2]

b) Yanqun says that osteoporosis is caused by smoking.
Suggest why this statement is **not** true.

_____ [1]

c) Which two nutrients might someone suffering from osteoporosis fail to absorb?

_____ and _____ [2]

d) What advice might you give to someone about lifestyle choices to prevent osteoporosis?

_____

_____ [2]

2 Read the information below and then answer the questions that follow.

Alcohol is produced by a process called **fermentation**. Yeast breaks down sugar into **alcohol** and **carbon dioxide**. It is important that no oxygen is present or the yeast will produce **ethanoic acid** – the chemical found in vinegar.

In commercial production, large steel fermentation tanks like the one in the picture are used. The tanks are sterilised before use; the ingredients are added and conditions inside the fermentation tanks are monitored carefully. The yeast will produce **energy as heat** and it is important that the temperature does not rise above 40°C. This method can be used for producing both wine and beer.

a) Name the two products produced by yeast in a fermentation reaction.

_____ and _____ [2]

b) Why is it important that the yeast ferments anaerobically?

_____

_____ [2]

c) Suggest one condition that needs to be monitored inside the fermentation tanks.

_____ [1]

d) What will happen if the temperature rises above 40°C?

_____ [1]

e) What might happen if the temperature falls too low?

_____ [1]

f) Suggest why the fermentation tanks are sterilised prior to use.

_____

_____ [2]

Total Marks _____ / 16

	Vocabulary Builder	Maths Skills	Testing Understanding	Working Scientifically	Science in Use
Total Marks	/ 35	/ 19	/ 41	/ 5	/ 16

15

Looking at Plants and Ecosystems

Vocabulary Builder

1 For each term, circle the word that makes the definition correct.

a) **Photosynthesis:**

The process by which plants produce **protein / glucose** [1]

b) **Algae:**

Single- or multi-celled **animals / plants** that are found in moist places [1]

c) **Reactants:**

The chemicals that we **start / finish** with in a chemical reaction [1]

d) **Ecosystem:**

All the animals and plants in an area and how they **breed / interact** with each other and the environment [1]

e) **Producer:**

The **animal / plant** at the start of a food chain [1]

2 Write the word that each of the statements below is defining.

a) The mass of living material .... [1]

b) Water loss in plants .... [1]

c) The major minerals used by plants and needed in larger quantities .... [1]

d) The role played by an organism within an ecosystem .... [1]

3 Fill in the spaces in the paragraph below about photosynthesis. The first letter of each missing word has been given.

Photosynthesis is the process by which leaves absorb **l** .... and **c** .... **d** .... to produce **g** ...., which is then used by the plant to produce other **m** .... such as **p** .... and complex **c** .... [6]

4 **a)** What word describes the part of the stem that transports water from the roots to the leaves? .... [1]

b) What word describes the part of the stem that transports glucose around the plant? .... [1]

c) What word describes the upper and lower layers of the leaf? .... [1]

d) What term describes the cells that open and close the stomata? .... [1]

e) What word describes the cell structures that contain chlorophyll? .... [1]

5 For each definition, choose the most appropriate term from the box below.

prey	carnivore	top predator	habitat	producer

a) The green plant at the start of a food chain .... [1]

b) An animal that feeds on other animals .... [1]

c) The animal at the end of a food chain .... [1]

d) An animal that gets eaten by another animal .... [1]

e) The place where an animal or plant lives .... [1]

6 The words below describe types of interdependence between organisms. Draw lines to match each word with its correct description.

Word   **Description**

Commensalism		One organism benefits from the relationship; the other is harmed
Mutualism		One organism benefits from the relationship; the other is not harmed
Parasitism		Both organisms benefit from the relationship

[2]

Total Marks .... / 27

1 The graph below shows the rate of photosynthesis of a plant on a summer day and a winter day.

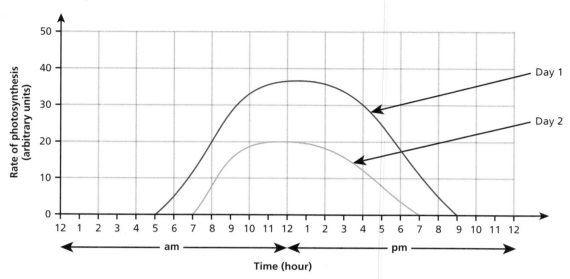

a) Suggest which day is the summer day. .... [1]

b) At midday on Day 1, what was the rate of photosynthesis? .... [1]

c) Over how many hours on Day 2 did the plant photosynthesise? .... [1]

d) Suggest two reasons why there is a difference in rate of photosynthesis on the two days.

....

....

....
[2]

2 Minerals are needed in small amounts to keep plants healthy.
The label below shows the ingredients of 'Grow All' fertiliser.
Draw a pie chart in the circle to show the composition of 'Grow All'.

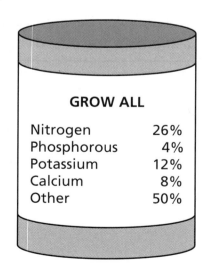

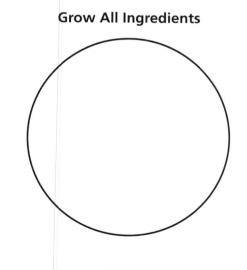

Grow All Ingredients

[4]

Total Marks .... / 9

Testing Understanding

1 **a)** Complete the word equation for photosynthesis.

.... + Water → .... + ... [3]

b) Explain how root hair cells are adapted to absorb water from the soil.

.... [1]

c) Which plant organ carries out photosynthesis? .... [1]

2 Look at the food chain below.

Leaf → Beetle → Fox

a) Name a producer in the food chain. .... [1]

b) Name a herbivore in the food chain. .... [1]

c) Name a consumer in the food chain. .... [1]

d) Leaves are also eaten by worms.

Birds eat worms and beetles.

Add this information to the food chain above to make a food web. [3]

e) Explain why all the energy in the leaf is not transferred to the fox.

....

....

.... [3]

3 Ground beetles are generalists and eat a variety of food including leaves and fungi.
Dung beetles eat only faeces.
What is the name given to organisms, such as the dung beetle, that have a narrow ecological niche?

.... [1]

4 The food web below is part of a farmland food web.

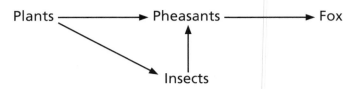

a) Name a secondary consumer in the above food web. _____ [1]

b) Use the idea of **bioaccumulation** to explain why harmful levels of pesticides might be found in a fox's body.

_____

_____

_____

_____

_____ [3]

c) Some farmers leave an area that they do not spray with pesticides around the edge of their fields.
How would this affect the number of pheasants in the area? Explain your answer.

_____

_____ [2]

d) In the 1940s, farmers used a pesticide called DDT.
This pesticide caused the egg shells of some birds to be thinner than usual.
Explain how this might affect the population of affected birds.

_____

_____ [2]

5 Plants living in the same environment will compete with each other.

a) Name two things that plants might compete for.

_____ and _____ [2]

b) Dandelions have a long tap root and large leaves.
Explain how these features make them good competitors.

_____

_____ [2]

6 Complete the passage below about energy transfer by filling in the missing words.

Energy from the _____ is the source of energy for all living things.

Plants convert some of this energy into glucose using the process of _____ .

This energy passes through the food chain but a large amount is _____ at

each stage of the chain. Energy is lost when animals _____ ,

_____ and _____ . In warm blooded animals it is also lost as

_____ .   [7]

7 In what conditions is transpiration in plants likely to be at its highest?

_____ and _____   [2]

8 Read the paragraph below about how leaves are adapted for photosynthesis.
Fill in the missing words.

The upper epidermis of the leaf is very thin to allow _____ to easily pass

through. The palisade layer has cells that contain many _____ and is where

_____ occurs. The spongy mesophyll layer has many air spaces to allow

_____ and _____ to move through the leaf.

_____ open and close the stomata on the underside of the leaf.   [6]

9 Monoculture is a practice where a single crop is grown in a vast area.
Explain why this is likely to reduce the number of natural insect pollinators in the area.

_____

_____   [2]

10 Name three physical factors that affect the environment an animal or plant may live in.

_____

_____   [3]

11 Jay says that the Sun is the source of all our food. Explain why he says this.

_____

_____

_____

_____   [5]

12 Complete the passage below by circling the correct terms in bold.

Chemosynthesis, unlike photosynthesis, does not require **light / nitrogen**. Chemosynthetic **bacteria / viruses** are found in deep sea communities. They use chemicals such as **ammonia / hydrogen sulfide** to get energy, which is then used to convert carbon dioxide into **glucose / nitrates**. Some chemosynthetic organisms are found living inside giant tubeworms. They get the chemicals they need from the tubeworms; the tubeworms use the sugars produced by them as food. This is an example of **parasitism / mutualism**. [5]

13 Plant fertilisers are often called NPK fertilisers. Each letter represents a mineral present in the fertiliser. What mineral does each letter represent?

N = ... P = ... K = ... [3]

14 The diagram below shows a cross section of a leaf.

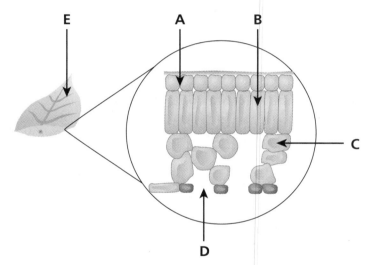

a) Which letters represent the following? You may need to use some letters more than once.

   i) A vein .... [1]

   ii) The upper epidermis .... [1]

   iii) Where gases pass into and out of the leaf .... [1]

   iv) The layer responsible for most photosynthesis .... [1]

   v) A stoma .... [1]

   vi) Palisade cells .... [1]

b) Some leaves have a waxy cuticle. What is the function of this?

   .... [1]

Total Marks .... / 67

1 A variegated leaf is one that has both green and white areas.

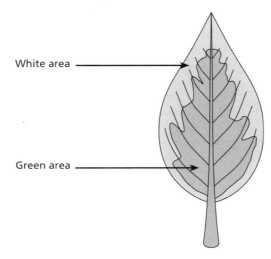

White area

Green area

a) What substance present in the leaf is responsible for the green colour?

_____ [1]

b) Sean tested the leaf to find out which areas contain starch.
   What is the name of the chemical he would use to test for the presence of starch?

_____ [1]

c) Label the leaf below after testing for starch, showing the colour you would expect
   each area to be.

A: _____

B: _____

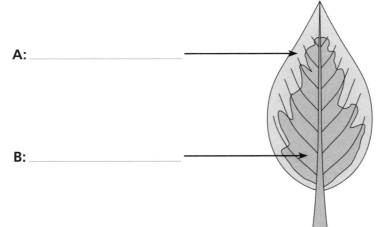

[2]

d) In what conditions should the plant be kept before testing the leaf for starch?

_____ [1]

e) What is the name of the process by which plants produce glucose?

_____ [1]

2 The apparatus below was set up to investigate photosynthesis of the pondweed Elodea. Students planned to count the number of bubbles produced by the Elodea in two minutes when the lamp was at different distances from the Elodea.

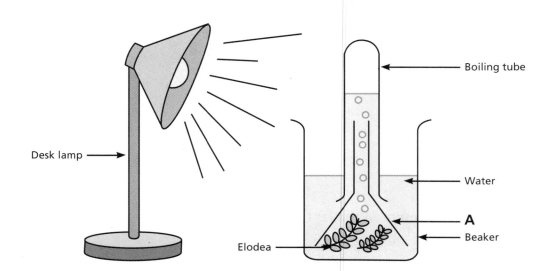

a) What is the name of the piece of equipment labelled **A**?

.... [1]

b) What is the **independent** variable in this experiment?

.... [1]

c) What is the **dependent** variable in this experiment?

.... [1]

d) Suggest two additional pieces of equipment, not shown in the diagram, that the students would need to carry out the experiment.

.... and .... [2]

e) Predict how the number of bubbles would change as the lamp was moved nearer to the Elodea.
Give a reason for your answer.

....

.... [2]

Total Marks .... / 13

1 Read the passage below about acid rain and then answer the questions that follow.

> **Air pollution** by gases such as sulfur dioxide **causes acid rain**, which is known to damage forests. Pine trees are most affected because the needles are bathed in acid droplets all year round; other trees drop their leaves. The pH of normal acid rain is around 4.2 but forests at high altitudes may be shrouded by clouds or fog for much of the time. The pH of cloud droplets averages 3.6. In a recent survey in Germany, it was found that the numbers of red spruce had decreased by 70% in the last 30 years, whilst the number of sugar maple trees had decreased by only 25%.
>
> Acid rain damages leaves because it **causes minerals such as calcium, magnesium and potassium to be lost** from them. Black horned pine borers are beetles that do not attack healthy trees but are attracted to dying, diseased or stressed trees. These beetles lay their eggs beneath the bark of the tree. The larvae that hatch feed on the phloem.

a) Name one gas that causes acid rain. _____ [1]

b) Suggest two possible reasons why numbers of red spruce have declined more than sugar maple trees in the past 30 years.

_____

_____ [2]

c) How do trees obtain minerals such as calcium and magnesium?

_____ [2]

d) Suggest why trees infested with black horned pine borers have an increased risk of dying.

_____

_____

_____ [2]

e) The food chain below is part of a food web found in the forest.

Conifer ⟶ Caterpillars ⟶ Red-winged blackbirds

How would the decline in conifer trees affect the population of red-winged blackbirds? Explain your answer.

_____

_____

_____ [2]

2 Read the following passage and then answer the questions that follow.

> **Pollinators** such as bees, birds and bats affect 35% of the world's crop production. These pollinators help spread the pollen needed for **fertilisation** of crops such as fruit, vegetables and nuts. However, pollinator **biodiversity** is rapidly decreasing due to modern farming methods. Research has shown that animal-pollinated crops tend to be high in vitamins and minerals, and suggests that falling numbers of pollinators will have a drastic effect on human nutrition.
>
> In Brazil, passion fruits are hand-pollinated by labourers, which has sent prices rocketing. It is feared that, as a result of this, many people will turn to less healthy foods such as fatty meats and sugars.
>
> Suggestions to improve pollinator biodiversity include reducing the amount of insecticides used and leaving borders of wild flowers and weeds around the edge of crop fields. Crops that tend to be less affected by decreasing numbers of pollinators include cereals, such as wheat, corn and rice.

a) What does the term **biodiversity** mean?

.... [1]

b) Why are animal-pollinated crops important in a healthy diet?

.... [1]

c) Suggest one consequence of people in Brazil turning to less healthy foods.

.... [1]

d) Suggest how leaving borders of wild flowers and weeds around the edge of crop fields will help to improve pollinator diversity.

....

.... [2]

e) Suggest a reason why crops such as wheat and rice are less affected by decreasing pollinator numbers.

.... [1]

Total Marks .... / 15

	Vocabulary Builder	Maths Skills	Testing Understanding	Working Scientifically	Science in Use
Total Marks	.... / 27	.... / 9	.... / 67	.... / 13	.... / 15

Chemistry   Explaining Physical Changes

Vocabulary Builder

1 Draw lines to match the following changes of state to their definitions.

Change of State	Definition
Melting	A solid turning directly into a gas (or a gas turning directly into a solid)
Freezing	A solid turning into a liquid
Boiling	A gas turning into a liquid
Condensing	A liquid turning into a solid
Subliming	A liquid turning into a gas

[4]

2 Which of the following statements about **diffusion** are true (**T**) and which are false (**F**)? Write **T** or **F** in the spaces provided.

a) Diffusion involves the random movement of particles. _____ [1]

b) During diffusion, particles move from an area of high concentration to an area of low concentration. _____ [1]

c) When the particles have spread out, the particles stop moving. _____ [1]

d) Ink spreading through water is an example of diffusion. _____ [1]

3 What does **malleable** mean?
Underline the correct definition.

A: Brittle

B: Can be hammered into shape

C: Can be drawn into extremely thin wires

D: Strong

[1]

4 Complete the following sentence using one of the words from the box.

air	nitrogen	steam	steel

_____ cannot be **compressed**. [1]

5 Underline the event that can be explained by **Brownian motion**.

    A: Water evaporating

    B: Ice floating on water

    C: Solid carbon dioxide turning into a gas upon heating

    D: The random motion of smoke particles trapped in air [1]

6 Which one of the diagrams below does **not** contain **molecules**?
Explain your answer.

A

B

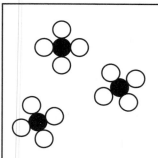

C

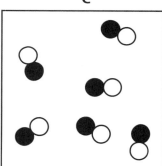

D

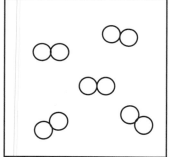

Diagram: _____ [1]

Explanation: _____ [1]

Total Marks _____ / 13

1 Sally placed solid X into a beaker and slowly heated it until it melted and then boiled.
She recorded the temperature every minute.
Her results are shown on the graph below.

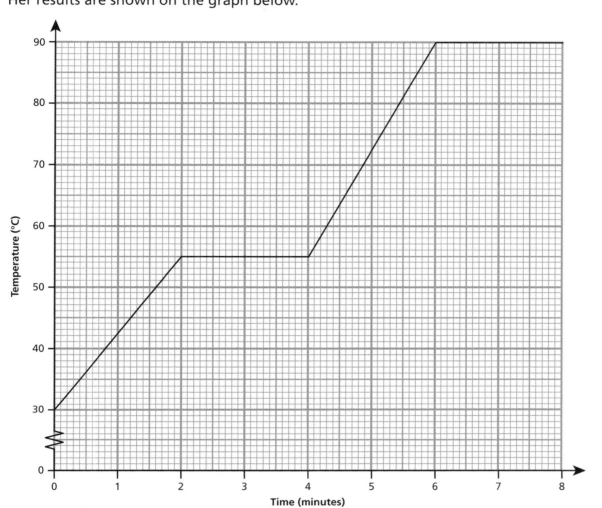

a) What is the melting point of solid X? .... [2]

b) What is the boiling point of solid X? .... [2]

c) How long did it take solid X to melt? Show your working.

....

....

.... [2]

2 Substance Y melts at −36°C and boils at 95°C.
At which one of the following temperatures will Y be a liquid?
Underline the correct option.

   A: −50°C   **B:** 0°C   **C:** 100°C   [1]

3 Density is calculated using the formula below.

$$\text{Density} = \frac{\text{Mass}}{\text{Volume}}$$

a) 10cm³ of water has a mass of 10.0g. Calculate the density of water and show your working.

....

.... g/cm³ [2]

b) 10cm³ of ice has a mass of 9.0g. Calculate the density of ice.

....

.... g/cm³ [2]

c) Using your answers to parts a) and b), explain why ice floats on water.

.... [2]

Total Marks .... / 13

Testing Understanding

1 A student was using a microscope to observe a pollen grain floating on a drop of cold water. He noticed that it appeared to be randomly moving and made a sketch to show how the pollen grain moved over a period of a minute. His sketch is shown below.

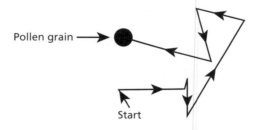

a) What name is given to the process by which the pollen grain randomly moves?

.... [1]

b) Explain why the pollen grain appears to be moving.

....

....

.... [2]

c) Explain what would happen to the movement of the pollen grain if it was placed on a drop of hot water.

.... [1]

d) Explain your answer to part **c)**.

....

....

....

.... [2]

2 The diagram below shows how solids, liquids and gases can be changed from one state of matter to another.

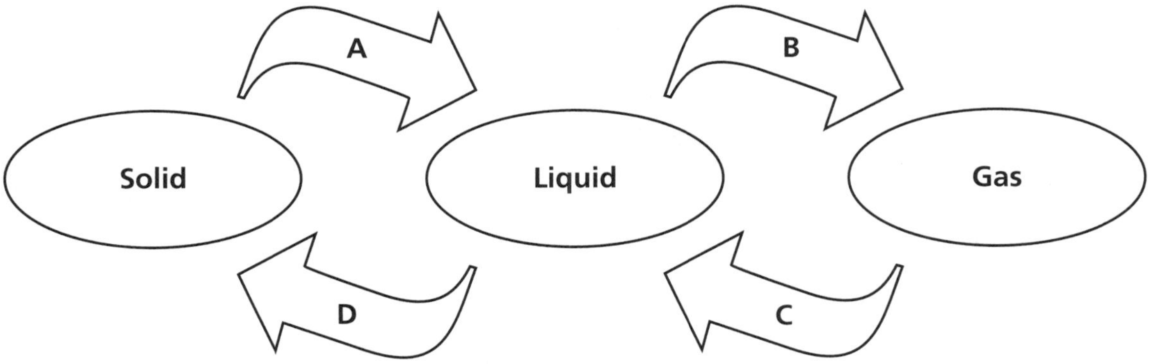

a) Choose the correct name from the list below for each of the changes of state.

boiling	freezing	melting	condensing

A: ....

B: ....

C: ....

D: .... [4]

b) What other word can be used to describe changing into a liquid from a gas?

.... [1]

c) A few substances change directly from a solid to a gas or from a gas to a solid. What name is given to this process?

.... [1]

3 In a fume cupboard, a teacher placed an upturned gas jar containing air onto a gas jar containing a few drops of liquid bromine. A brown vapour immediately started to form at the bottom of the gas jar, as shown in the image below. The gas jars were then left until the next lesson.

Brown vapour

a) What caused the formation of the brown vapour?

.... [1]

b) Circle the letter of the diagram below that most accurately represents the arrangement of particles in liquid bromine.

A   B   C

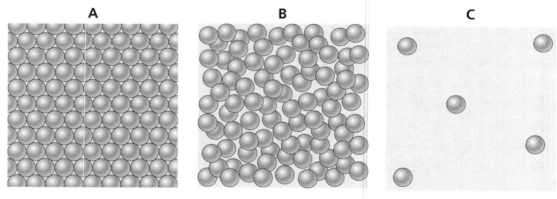

[1]

c) Describe what the class would have observed when they looked at the gas jars during the next lesson.

....

.... [1]

d) What name is given to the process by which particles move from an area of high concentration to an area of low concentration?

.... [1]

e) In your own words, explain how this process occurs.

_____

_____ [2]

④ Potassium permanganate is a purple dye. In an experiment, a teacher placed some potassium permanganate into a beaker of water. The diagram below shows in terms of particles what happens to the dye after it has been added to the water.

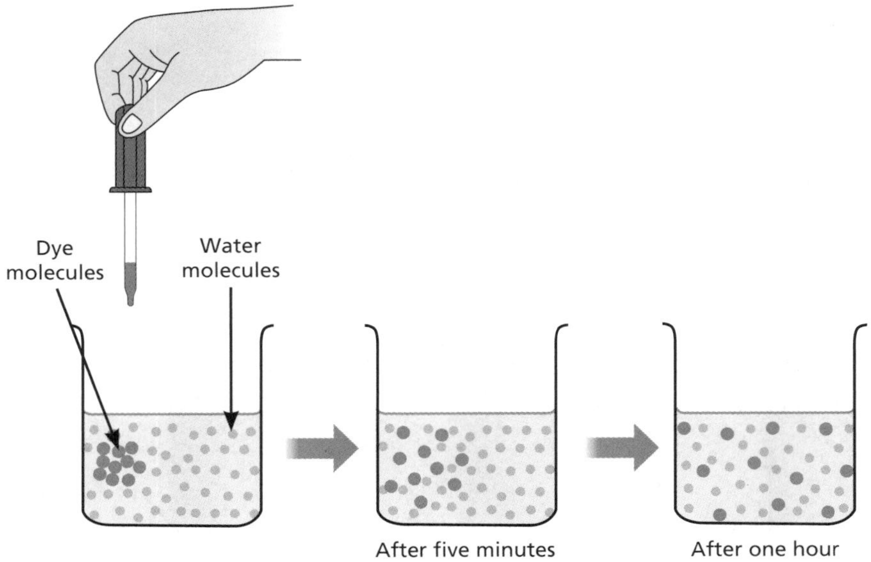

Dye molecules   Water molecules   After five minutes   After one hour

a) Describe the appearance of the beaker of water after five minutes and after one hour.

_____

_____

_____ [2]

b) What name is given to the process shown above? _____ [1]

c) In terms of particles, explain how this process occurs.

_____

_____

_____ [2]

d) Describe the appearance of the beaker after two hours.
Explain your answer.

_____

_____

_____ [2]

5 State whether each of the following statements about solids, liquids and gases is true (**T**) or false (**F**). Write **T** or **F** in the spaces provided.

a) Solids have a fixed shape and volume. .... [1]

b) Water is denser than ice. .... [1]

c) Liquids have a fixed shape. .... [1]

d) Gases have a fixed volume. .... [1]

e) Particles in a solid are randomly arranged. .... [1]

f) Particles in a gas move about very quickly. .... [1]

g) Particles in a solid have more energy than particles in a gas. .... [1]

6 A student placed some candle wax in a beaker. He slowly heated the beaker and, using a thermometer, he recorded the temperature every minute and plotted his results on a graph. His results are shown below.

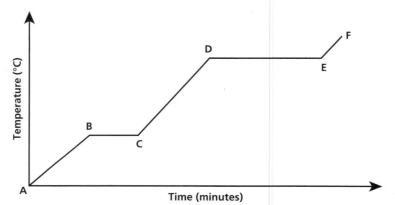

When answering the following questions, use two letters to represent the part of the graph you are referring to (e.g. **AB** means the part of the graph that starts at **A** and finishes at **B**).

a) Which section of the graph refers to the melting point? .... [1]

b) In which section of the graph were the particles arranged as shown below? .... [1]

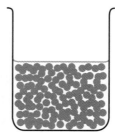

c) Which sections of the graph show that the energy of the particles is increasing?

_____ [3]

d) What state of matter is the substance between E and F? _____ [1]

e) Name the process occurring between D and E. _____ [1]

Total Marks _____ / 39

Working Scientifically

1 Sophie was investigating the three states of matter of water. She placed some ice from the freezer into a beaker and then placed the beaker on a balance as shown in the diagram below. Sophie returned to the beaker when all of the ice had melted.

a) The empty beaker had a mass of 82g.
Calculate the mass of ice in the beaker.

_____ [2]

b) In the box, show the arrangement of the particles in ice.
Use ○ to represent each particle.

[2]

c) What change of state takes place when melting occurs?

_____ [1]

d) Explain what happens to the **energy** and **movement** of the particles when melting occurs.

_____

_____ [2]

e) What will be the mass reading on the balance shown on page 35 when the ice has melted? Explain your answer.

_____

_____ [2]

f) A few days later, Sophie placed the beaker on a balance and recorded the mass as 167.52g.
Explain why the reading had gone down.

_____ [1]

Total Marks _____ / 10

Science in Use

1 Read the passage about ice and then answer the questions that follow.

Most materials are **denser** when they are solid compared to when they are liquid. Ice is unusual in that the solid form of water floats on liquid water. This is why we can see icebergs and ice cubes floating on the surface of water.

Scientists are concerned that as the atmosphere warms up (a process called **global warming**) sea levels will rise.

Answers

Getting the Energy Your Body Needs

Pages 4–15

Vocabulary Builder

1. Aerobic respiration – Respiration using oxygen; Mitochondria – Where respiration takes place; Anaerobic respiration – Respiration in the absence of oxygen; Lactic acid – Substance that builds up in muscles during anaerobic respiration; Fermentation – Process that converts sugar into alcohol
   [All correct = 4 marks; 3 correct = 3 marks; 2 correct = 2 marks; 1 correct = 1 mark]
2. a) Vertebrate **[1]**
   b) Bone marrow **[1]**
   c) Collagen **[1]**
   d) Cardiac **[1]**
   e) Mitochondria **[1]**
   f) Glycogen **[1]**
   g) (Anabolic) steroids **[1]**
3. heart, lungs **(in either order) [2]**; skull **[1]**; support **[1]**; move **[1]**; blood cells **[1]**; joints **[1]**; cartilage **[1]**; synovial fluid **[1]**; ligaments **[1]**
4. a) Antagonistic **[1]**
   b) i) quadriceps **[1]**
      ii) deltoid **[1]**
5. Hinge Joints – elbow **[1]**; knee **[1]**; finger **[1]**
   Ball and Socket Joints – shoulder **[1]**; hip **[1]**
   Gliding / Pivot Joints – neck **[1]**; wrist **[1]**
6. a) sternum **[1]**; because the others are all bones in the arm **[1]**
   b) scapula **[1]**; because the others are all bones in the leg **[1]**

Maths Skills

1. a)
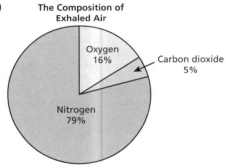
The Composition of Exhaled Air

Oxygen 16%
Carbon dioxide 5%
Nitrogen 79%

   [3 marks for all sectors correct; –1 for each error]
   b) **B:** About a quarter of the oxygen inhaled is used in respiration. **[1]**

2. a)–c)   If the points form a curve, do not use a ruler to join them.

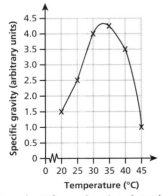

   [9 marks: 1 for each point plotted correctly; 1 for each axis correctly labelled; 1 for joining the points with a curve]

---

   d) **Accept:** any answer between 3.2 and 3.4 **[1]**
   e) 25°C **[1]**
   f) 35°C **[1]**
3. a) Points joined with a curve **[1]**
   b) **Accept:** any answer between 4.8 and 5.2mmol/l **[1]**
   c) 3.8m/s **[1]**

Testing Understanding

1. a) Aerobic **[1]**
   b) Anaerobic **[1]**
   c) Aerobic **[1]**
   d) Aerobic **[1]**
   e) Anaerobic **[1]**
2. Glucose **[1]** → Lactic acid **[1]** + Energy **[1]**
3. a) **Left of arrow:** Oxygen **[1]**
      Right of arrow: Carbon dioxide **[1]**
   b) In the blood **[1]**
   c) Mitochondria **[1]**
4. a) Fermentation **[1]**

   b)   Always make sure reactants are on the left side of the arrow and products are on the right side. The order of products is not important.

      Glucose **[1]** → Ethanol **[1]** + Carbon dioxide **[1]** + Energy **[1]**
5. a) Biceps **[1]**
   b) Triceps **[1]**
   c) Muscle A (biceps) contracts **[1]**
      Muscle B (triceps) relaxes **[1]**
   d) Tendons **[1]**
   e) taking anabolic steroids **[1]**; regular exercise **[1]**
6. **In either order:** shorter **[1]**; fatter **[1]**
7. a) **A:** Ball and socket **[1]**
      B: Hinge **[1]**
   b) **Accept either:** Neck; Wrist **[1]**
8. **B:** The air that is breathed in contains more oxygen and less carbon dioxide than the air that is breathed out. **[1]**
9. a) **D:** When one muscle contracts, the other relaxes **[1]**
   b) Muscles require more energy **[1]**; therefore, more oxygen and glucose are needed **[1]**. These are carried in the bloodstream **[1]** so the heart beats faster to increase blood flow **[1]**.
10. a) **A:** Skull **[1]**
       B: Sternum **[1]**
       C: Ribs **[1]**
       D: Pelvis **[1]**
       E: Femur **[1]**
    b) Hinge **[1]**

Working Scientifically

1. a) To detect presence of carbon dioxide **[1]**
   b) Tube B **[1]**
   c) It should turn cloudy / milky **[1]**
   d) Control **[1]**
   e) The woodlice may become distressed / die at this temperature **[1]**

Science in Use

1. a) Bones do not begin to lose their density **[1]** until the age of 35 **[1]**
   b) It only increases the risk **[1]**
   c) **In either order:** Calcium **[1]**; Vitamin D **[1]**
   d) **Accept any two from:** Do not smoke; Do not drink heavily; Eat a diet rich in calcium and vitamin D **[2]**
2. a) **In either order:** Carbon dioxide **[1]**; Alcohol **[1]**
   b) If oxygen is present it will produce ethanoic acid instead of alcohol **[1]**; It will taste horrible **[1]**
   c) **Accept any one from:** Temperature; Oxygen levels; pH **[1]**
   d) The yeast will die **[1]**

e) **Accept either:** Fermentation will be too slow; Not enough alcohol will be produced [1]
f) To kill unwanted bacteria or yeasts [1], which would spoil the beer or wine [1]

Looking at Plants and Ecosystems

Pages 16–26

Vocabulary Builder
1. a) glucose [1]
    b) plants [1]
    c) start [1]
    d) interact [1]
    e) plant [1]
2. a) Biomass [1]
    b) Transpiration [1]
    c) Macroelements [1]
    d) Niche [1]
3. light [1]; carbon dioxide [1]; glucose [1]; molecules **or** materials [1]; proteins [1]; carbohydrates [1]
4. a) Xylem [1]
    b) Phloem [1]
    c) Epidermis [1]
    d) Guard cells [1]
    e) Chloroplasts [1]
5. a) Producer [1]
    b) Carnivore [1]
    c) Top predator [1]
    d) Prey [1]
    e) Habitat [1]
6. Commensalism – One organism benefits from the relationship; the other is not harmed
    Mutualism – Both organisms benefit from the relationship
    Parasitism – One organism benefits from the relationship; the other is harmed
    [2 marks for all three correct; 1 mark for one correct]

Maths Skills
1. a) Day 1 [1]
    b) **Accept:** any answer between 35 and 38 (arbitrary units) [1]
    c) 12 hours [1]
    d) **In either order:** Difference in temperatures [1]; Difference in amount of sunlight [1]

2.
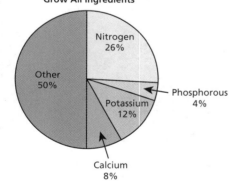

Grow All Ingredients

Nitrogen 26%
Other 50%
Phosphorous 4%
Potassium 12%
Calcium 8%

[4 marks for all sectors correct; –1 for each error]

Testing Understanding
1. a) **Left of arrow:** Carbon dioxide [1]
        Right of arrow in either order: Glucose [1] + Oxygen [1]
    b) They have a large surface area [1]
    c) Leaf [1]
2. a) Leaf [1]
    b) Beetle [1]
    c) **Accept either:** Beetle; Fox [1]

d) 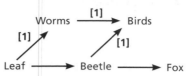 Make sure the arrows point in the correct direction. Remember, the arrows in a food chain / web show the direction of energy transfer. You can then work out what eats what using that information.

Worms —[1]→ Birds
[1]↗ ↗[1]
Leaf —→ Beetle —[1]→ Fox

e) Energy in the beetle is lost in movement / respiration [1] and excretion [1]. The beetle may not eat all the leaf [1].
3. Specialists [1]
4. a) **Accept either:** Fox; Pheasants [1]

    b) It is important to get across the point that the levels accumulate over time.

    Insects will consume a small amount of pesticides that will not harm them [1]. Pheasants consume many insects and will have higher levels of pesticides [1]. The fox consumes many pheasants so pesticide levels build up to harmful levels over time [1].
    c) It will increase the number of pheasants [1] because there would be more insects to feed on [1]
    d) Numbers would decrease [1] because eggs would break too early, before they're ready to hatch [1]
5. a) **Accept any two from:** Light; Space; Water; Minerals [2]
    b) Large leaves catch a lot of sunlight [1]; Long tap root provides anchorage / stops them from being pulled up **or** Long tap root enables access to water / minerals that other plants may not reach [1]
6. Sun [1]; photosynthesis [1]; lost [1]; respire, move, excrete **(in any order)** [3]; heat [1]
7. **In either order:** Hot [1]; Dry [1]
8. sunlight [1]; chloroplasts [1]; photosynthesis [1]; oxygen, carbon dioxide **(in either order)** [2]; guard cells [1]
9. Crops flower for only a short time [1] and there are no other plants for insects to feed on at other times [1]
10. **In any order:** Temperature [1]; Light intensity [1]; Amount of water [1] **(accept other suitable factors)**
11. Sun is needed for plants to photosynthesise [1] in order to grow [1]. Plants are food for herbivores [1; **accept a named example, e.g. cows**]. Herbivores are then eaten by carnivores / omnivores [1; **accept a named example, e.g. hens / humans**]. We eat plants, herbivores and omnivores / carnivores [1].
12. light [1]; bacteria [1]; hydrogen sulfide [1]; glucose [1]; mutualism [1]
13. N = Nitrogen [1]
    P = Phosphorous [1]
    K = Potassium [1]
14. a) i) E [1]
        ii) A [1]
        iii) D [1]
        iv) B [1]
        v) D [1]
        vi) B [1]
    b) It helps to reduce water loss from the leaf [1]

Working Scientifically
1. a) Chlorophyll [1]
    b) Iodine [1]
    c) A: Brown [1]
       B: Black [1]
    d) In the light [1]
    e) Photosynthesis [1]
2. a) Filter funnel [1]
    b) The distance of the lamp from the Elodea [1]
    c) The number of bubbles produced in two minutes [1]
    d) **In either order:** Stopwatch [1]; Ruler [1]
    e) It would increase [1], because the increase in light will increase the rate of photosynthesis [1]

Science in Use

1. a) **Accept either:** Sulfur dioxide; Nitrous oxide [1]
   b) Red spruce do not drop leaves, whereas maple trees do [1];
      Red spruce grow at high altitude, maple trees do not [1]
   c) From the soil [1] through their roots [1]
   d) The pine borers feed on the nutrients / glucose in the phloem [1]; therefore, the tree does not get enough nutrients to grow / make other molecules [1]
   e) The numbers of red-winged blackbirds would decrease [1], because there would be fewer caterpillars to feed on [1]

2. a) The variety of living organisms found in an area [1]
   b) They are high in vitamins and minerals [1]
   c) **Accept any one from:** More people will be overweight; More people will have an increased risk of heart disease; More people will have an increased risk of diabetes [1]
   d) There will be a wider variety of plants for the pollinating insects to feed on [1] for a longer period of time [1]
   e) They do not rely on animals to pollinate them / they are wind-pollinated [1]

Explaining Physical Changes

Pages 27–38

Vocabulary Builder

1. These are important words that we frequently use in science – make sure you know their meaning.

   Melting – A solid turning into a liquid; Freezing – A liquid turning into a solid; Boiling – A liquid turning into a gas; Condensing – A gas turning into a liquid; Subliming – A solid turning directly into a gas (or a gas turning directly into a solid) **[All correct = 4 marks; 3 correct = 3 marks; 2 correct = 2 marks; 1 correct = 1 mark]**

2. a) T [1]
   b) T [1]
   c) F [1]
   d) T [1]
3. **B:** Can be hammered into shape [1]
4. Steel [1]
5. **D:** The random motion of smoke particles trapped in air [1]

6. Remember that molecules consist of atoms chemically joined together.

   Diagram A [1]
   This is the only diagram containing single atoms [1]

Maths Skills

1. When taking readings from a graph, use a ruler to draw from the plotted line to the axes.

   a) 55°C **[2 marks: 1 for the numerical answer; 1 for the unit]**
   b) 90°C **[2 marks: 1 for the numerical answer; 1 for the unit]**
   c) 2 minutes [1]
      There are 20 small squares on the horizontal section of the graph at 55°C, which corresponds to 2 minutes **or** Draw a line down to the x-axis from the start of the horizontal line at 55°C and another line down to the x-axis from the end of the horizontal line at 55°C; the difference on the x-axis between those two lines is 2 minutes [1]

2. **B:** 0°C [1]
3. a) 10.0(g) [1] ÷ 10(cm³) = 1.0g/cm³ [1]
   b) 9.0(g) [1] ÷ 10(cm³) = 0.90g/cm³ [1]

   c) Less dense materials float on more dense materials.

      Ice is less dense [1] than water [1]

Testing Understanding

1. a) Brownian motion [1]
   b) The water particles are constantly moving [1] and colliding with the pollen grain, causing it to move [1]
   c) It would move more quickly [1]
   d) In chemistry, almost everything happens faster at higher temperatures.

      The water particles have more energy when they are hot [1] and so will collide more quickly with the pollen grain [1]

2. a) **A:** Melting [1]
      B: Boiling [1]
      C: Condensing [1]
      D: Freezing [1]
   b) Evaporating [1]
   c) Subliming / sublimation [1]

3. a) The liquid bromine evaporated [1]
   b) B [1]
   c) The brown colour would have spread out equally into the two gas jars [1]
   d) Diffusion [1]

   e) Brownian motion and diffusion both involve the movement of tiny particles that are too small to see.

      The particles are in constant random motion [1] and, as they move, they bump into air particles and spread out [1]

4. a) After five minutes, there will be purple colour around where the dye was added [1]. After one hour, the purple colour will have spread throughout the beaker [1].
   b) Diffusion [1]
   c) Particles are in constant random motion [1] and, as they move, they bump into water particles and spread out [1]

   d) Even though the colour will not look like it is changing, the particles will still be moving.

      The colour will still be evenly spread throughout the beaker [1]. Diffusion still occurs, but the particles remain evenly spread throughout the beaker [1].

5. a) T [1]
   b) T [1]
   c) F [1]
   d) F [1]
   e) F [1]
   f) T [1]
   g) F [1]
6. a) BC [1]
   b) CD [1]
   c) AB [1]; CD [1]; EF [1]
   d) Gas [1]
   e) Boiling [1]

Working Scientifically

1. a) 168.65 – 82 [1] = 86.65g [1; full marks only if correct units are included]

   b)

   [2 marks: 1 for particles close together (without gaps); 1 for the regular arrangement / pattern]
   c) Solid turns to liquid [1]
   d) The particles gain energy [1] and vibrate / move more [1]
   e) 168.65g [1]; The mass does not change when melting occurs as the same amount of matter is present [1]
   f) Some of the water had evaporated [1]

Science in Use

1. a) Water is one of the few substances where the solid form is less dense than the liquid form.

    The solid form of a substance is usually denser than the liquid form [1]. This means that a solid substance would be expected to sink when added to its liquid state [1].

    b) They will melt [1]

    c) It is a reversible process [1]. Melting is a physical change and the reverse of melting is freezing [1].

    d) Lots of ice is above sea level, so if this melts the amount of water in the oceans will increase.

    They could rise [1]. If the atmosphere continues to warm up then the ice (at the north and south poles) could melt and this will cause the sea levels to rise [1].

2. a) Liquid turns to gas [1]
    b) Energy increases / particles gain more energy [1]
    c) Latent heat [1]
    d) i) The liquid gains energy, which causes it to evaporate [1]
       ii) From the skin / body [1]
       iii) Diffusion [1]

Explaining Chemical Changes

Pages 39–48

Vocabulary Builder
1. a) Acid [1]
    b) Salt [1]
    c) Alkali [1]
    d) Salt [1]
2. **D:** Has a pH of more than 7 [1]
3. a) T [1]
    b) F [1]
    c) T [1]
    d) T [1]
4. **D:** Water mixing with sulfuric acid [1]
5. Speed up chemical reactions [1] without being used up [1]
6. **B:** Petrol reacting with oxygen [1]
7. acids [1]; hydrogen [1]; 7 [1]; hydroxide [1]; 7 [1]; react [1]; soluble [1]; salt [1]

8. Acids, alkalis and salts are words that you will meet many times in your study of chemistry.

    C: Acid + Water → Salt + Alkali [1]

Maths Skills
1. a) 163.8g [**2 marks: 1 for the numerical answer; 1 for the unit**]
    b) 161.4g [**2 marks: 1 for the numerical answer; 1 for the unit**]
    c) 163.8 – 161.4 [1] = 2.4g [**1; full marks only if correct units are included**]
    d) The gas can escape into the atmosphere during the reaction, which is why the reading on the balance goes down during the experiment.

    After 15 minutes [1]. After this time the line is horizontal, meaning that the mass was no longer changing [1].
2. a) The manufacture of fertilisers [1]
    b) When estimating from a pie chart, you will always be allowed a range of answers as it can be difficult to get the exact answer.

    Accept: any value between 5% and 15% [1]

Testing Understanding
1. a) Hydrogen [1]
    b) Zinc [1]
    c) Water [1]
2. **In either position:** Oxygen / Air [1]; Heat [1]

3. The colours of universal indicator are the same as the colours of the rainbow.

    a) Stomach acid [1]
    b) Household ammonia [1]
    c) Water [1]
    d) **Accept either:** Baking soda; Household ammonia [1]
4. **D:** A reversible reaction [1]
5. a) **Accept any one from:** Measuring cylinder; Pipette; Burette [1]
    b) Indicator (**can give a specific name, e.g. litmus, universal indicator**) [1]
    c) The indicator would have changed colour to show an acid was present (**allow any indicator with correct colour change, e.g. litmus turns red, universal indicator turns red / orange / yellow, phenolphthalein turns colourless**) [1]
    d) Magnesium chloride [1]
    e) Add a lighted splint to a test tube containing the gas [1]. If hydrogen is present then you will hear a squeaky pop [1].

6. Hydrochloric acid always forms chloride salts, sulfuric acid forms sulfate salts and nitric acid forms nitrate salts.

    Iron chloride [1]; Calcium chloride [1]
    Iron sulfate [1]; Calcium sulfate [1]
    Iron nitrate [1]; Calcium nitrate [1]

7. The name of the metal in the alkali always forms the first part of the name of salts.

    Calcium chloride [1]; Aluminium chloride [1]
    Calcium sulfate [1]; Aluminium sulfate [1]
    Calcium nitrate [1]; Aluminium nitrate [1]
8. a) Atoms chemically joined (bonded) together [1]
    b) That two atoms of oxygen are present in a molecule of oxygen [1]
    c) Carbon (or C) [1] and oxygen (or O_2) [1]
    d) Carbon dioxide (or CO_2) [1]
    e) Carbon + Oxygen [1] → Carbon dioxide [1]

Working Scientifically
1. a) A chemical that can be used to determine whether a substance is an acid or alkali [1]
    b) Because it is safer to work with the cooled liquid [1]
    c) i) Red [1]
           Green [1]
       ii) If there is no colour change / the red cabbage indicator stays purple [1]

Science in Use
1. a) Neutralisation [1]
    b) To make it a fair test [1]. The results would not be comparable if different amounts of acid were used each time [1].
    c) Red [1]
    d) Because all of the acid had been neutralised [1] so there was no more acid to react with the indigestion tablets [1]
    e) If fewer tablets are needed to neutralise the acid, it means that each tablet neutralises more acid.

    Bellysoothe [1], as this required fewer tablets to neutralise the acid [1]

2. A titration is an experiment that reacts an acid with an alkali very accurately.

    a) So that we know the water is safe to drink [1], as too much acid or alkali can be dangerous [1]
    b) It accurately records the volume of liquid [1]; You can add liquid one drop at a time [1]
    c) The indicator will change colour [1]

Exploring Contact and Non-Contact Forces

Pages 49–58

Vocabulary Builder

1. insulating **[1]**; electrically **[1]**; discharged **[1]**; static **[1]**; voltage **[1]**
2. electrons **[1]**; negative **[1]**; positive **[1]**; loses **[1]**; attractive **[1]**; unlike **[1]**
3. a) centre **[1]**
   b) attracted to **[1]**; gravitational **[1]**
   c) weight **[1]**
   d) 10 **[1]**
4. a) F **[1]**
   b) T **[1]**
   c) F **[1]**
5. a) Altimeters **[1]**
   b) Barometers **[1]**
   c) Pascals **(accept hectopascals, atmospheres, millibars, mm Hg)** **[1]**
   d) It becomes lower (falls) **[1]**
   e) Surface winds are formed by the flow of air **[1]** from high pressure to low pressure **[1]**
6. a) gravitational field **[1]**
   b) friction **[1]**
   c) gravitational field **[1]**
   d) gravitational field **[1]**
   e) gravitational field **[1]**

Maths Skills

1. Pressure (N/m^2 or Pa) = $\frac{\text{Force (N)}}{\text{Area (m}^2)}$ or P = $\frac{F}{A}$ (Pa or N/m^2)
   [4 marks: 2 for correct expression; 2 for correct units; −1 for each error or omission]
2. Using P = $\frac{F}{A}$, P = 20 000 ÷ 16 **[1]** = 1250Pa **[2 marks: 1 for correct value; 1 for correct units]**

3. Change mass into weight (force).

   65kg = (65 × 10) = 650N **[1]**
   Pressure = 650 ÷ 0.32 **[1]** = 2031.25 ≈ 2000N/m^2 or 2000Pa **[1]**

4. a) A hectopascal (hPa) is equal to 100 pascals.

   100 000Pa = 1000 × 100Pa **[1]** = 1000hPa **[1]**

   b) Rearrange the formula to find force in terms of pressure and area.

   Force = Pressure × Area (or F = P × A) **[1]**
   = 100 000 × 1.4 = 140 000N **[1]**

   c) Weight = Mass × Gravitational field strength

   Mass = 140 000 ÷ 10 **[1]** = 14 000kg **[1]**

Testing Understanding

1. a) **Accept any two from:** Electric force; Magnetic force; Gravitational force **[2]**
   b) Frictional force **[1]**
2. Gaining electrons – Negatively charged **[1]**
   Losing electrons – Positively charged **[1]**
3. a)

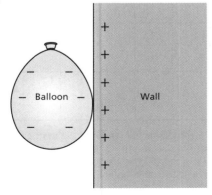

   [2 marks: 1 for correctly labelling the balloon; 1 for correctly labelling the wall]

   b) The balloon becomes negatively charged **[1]**. The negative charge on the balloon repels the electrons on the surface of the wall **[1]**. This leaves the wall with a net positive charge **[1]**. The negatively charged balloon is attracted to the positively charged wall **[1]**.
4. Electric field (lines) **[1]**
5. a) 1.7N/kg **[1]**
   b) 25.0N/kg **[1]**
6. The mass of the planet or moon **[1]** and its size **[1]**
7. a) T **[1]**
   b) T **[1]**
   c) F **[1]**
   d) T **[1]**
   e) F **[1]**
8. a) The large area of the snow shoes **[1]** results in lower pressure, which stops you sinking into soft snow **[1]**
   b) Air pressure falls with height gained **[1]**; need to maintain balanced pressure inside and outside of the aeroplane **[1]**
   c) Pressure increases with depth **[1]**, so strong suits are needed to withstand high pressure; heavy suits / weights are needed to take the divers to lower depths **[1]**

Working Scientifically

1. a)

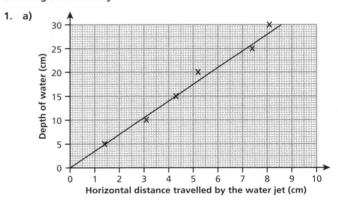

   Horizontal distance travelled by the water jet (cm)

   [4 marks: 3 for points plotted correctly, −1 for each error or omission; 1 for best line drawn using a ruler]
   b) Difference in pressure **(accept pressure)** **[1]**
   c) There is a linear relationship **[1]**. Depth of water (h) is directly proportional to the horizontal distance travelled by the water jet (l); or $h \propto l$. **[1]**
   d) P = 1000 × 10 × h **[1]** = 1000 × 10 × 0.30 **[1]** = 3000Pa **[1]**
   e) Density (of water) **[1]**
   f) i) P = 1000 × g × h = 1000 × 10 × 30 **[1]** = 300 000Pa = 300kPa **[1]**
      ii) Pressure at 30m = 300kPa = 3 × atmospheric pressure (i.e. 1 atmosphere/10m) **[1]**

Science in Use

1. a) N/kg **[1]**
   b) i) T **[1]**
      ii) T **[1]**
      iii) F **[1]**
      iv) T **[1]**
      v) T **[1]**
      vi) F **[1]**
   c) i) Weight = Mass × Gravitational field strength
         = 300 × 10 **[1]** = 3000N **[1]**
      ii) Weight = Mass × Gravitational field strength
         = 300 × 1.0 **[1]** = 300N **[1]**
      iii) Weight = Mass × Gravitational field strength
         = 300 × $\frac{10}{6}$ **[1]** = 500N **[1]**
      iv) At 18 000km above the Earth, the Earth's gravitational field strength is weak and the Moon's influence is even weaker **[1]**; As the vehicle approaches the Moon, the Moon's gravitational field becomes stronger and more dominant **[1]**

d) i) F [1]
    ii) T [1]
    iii) T [1]
    iv) F [1]
    v) T [1]
e) Initially, the Sun's weak gravitational field [1]; Eventually, the gravitational field of any other large objects, e.g. other stars [1]

Magnetism and Electricity

Pages 59–71

Vocabulary Builder
1. a) resistance; resistance [1; both answers are needed]
   b) Conductors: copper, iron, aluminium
      Insulators: plastic, rubber, polystyrene, wood, paint
      [4 marks: –1 for each error or omission]
2. iron [1]; domains [1]; N-pole, S-pole (in either order) [2]; repel [1]; attract [1]
3. current [1]; electrons [1]; positive [1]; negative [1]; amperes [1]; an ammeter [1]
4. a) A permanent magnet is a magnet that possesses its own magnetic field [1]. A temporary magnet is only magnetic when in contact with something else that is magnetic [1].
   b) Electromagnets are special types of temporary magnets [1]; they only become magnetic when a current is passed through them [1]. **Accept any two applications from:** Electric bell; Circuit breaker; Eye surgery; Loudspeakers; Computer hard drives; Separating magnetic / non-magnetic materials [2]
5. chemical [1]; electrons [1]; volts [1]; hot [1]
6. a) **A:** Solid (iron) core [1]
      B: Liquid core (of iron and nickel) [1]
      C: Mantle [1]
      D: Crust [1]
   b) The spinning Earth causes the liquid core to spin [1]; this causes charged particles to move, forming small currents [1]. These small currents create magnetic field domains [1]. Domains line up to produce a weak magnetic field [1].

Maths Skills
1. Using $R = \frac{V}{I}$, R = 230 ÷ 5 [1] = 46Ω [1; full marks only if correct units are included]
2. Rearranging $R = \frac{V}{I}$ gives $I = \frac{V}{R}$ [1] = 230 ÷ 23 [1] = 10A [1; full marks only if correct units are included]
3. a) Find the total resistance in the circuit.

   Total resistance = 2 × 3Ω = 6Ω [1]; $I = \frac{V}{R}$ = 6 ÷ 6 [1] = 1A [1; full marks only if correct units are included]

   b) The potential difference is shared equally between the two light bulbs.

   V = I × R [1] = 1 × 3 = 3V [1; full marks only if correct units are included]

   c) The potential difference is now proportional to the resistance.

   V = I × R = 1 × 4 [1] = 4V [1; full marks only if correct units are included]
4. a) 6V across each of the three bulbs [1]

   b) The total current is shared between all three bulbs.

   Using $I = \frac{V}{R}$ for one bulb gives I = 6 ÷ 2 [1] = 3A [1]; Current across each bulb = 3A [1]

Testing Understanding
1. The two poles will repel each other [1]
2. a) **A:** Series [1]
      B: Parallel [1]
      C: Parallel [1]

D: Parallel [1]
E: Series [1]
F: Parallel [1]
b) i) It would stay the same brightness (or similar, e.g. no change / no effect) [1]
   ii) It would go out (as the circuit is broken) [1]
3. a) F [1]
   b) T [1]
   c) F [1]
   d) T [1]
4. a) The wire will remain motionless / not move [1]
   b) The wire will move [1]
   c) There are two magnetic fields: one from the magnet [1] and one generated by the current in the wire [1]. The two fields will attract / repel [1], causing the wire to move [1].
   d) The loop of wire will rotate [1]. It is called the motor effect [1].
5. Increase the current [1]; Increase the strength of the magnet / obtain a stronger magnet using a different magnetic material [1]; Increase the number of turns of wire or coils [1]
6. a) F [1]
   b) T [1]
   c) T [1]
   d) T [1]
7. The compass needle always points towards the magnetic north pole [1] and this is close to the geographical north pole in the UK [1]; hence, you always (roughly) know your direction of travel [1] (or similar)
8. a) **Accept any two viable methods, e.g.:** Stroking it with a permanent magnet; Placing it next to a permanent magnet; Placing it in a coil of wire that has a direct current passing through it [2]
   b) **Accept any two viable methods, e.g.:** Heating; Hammering; Dropping; Using an alternating current [2]

Working Scientifically
1. a) series [1]; parallel [1]
   b)

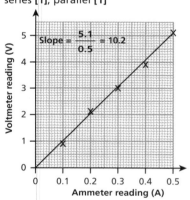

[4 marks: 3 for points plotted correctly, –1 for each error or omission; 1 for best line drawn using a ruler]

   c)

Ammeter Reading (A)	0.1	0.2	0.3	0.4	0.5	
Voltmeter Reading (V)	0.9	2.1	3.0	3.9	5.1	
Resistance (Ω)	9.0	10.5	10.0	9.8	10.2	Average: 9.9

[3 marks for 5–6 resistance values correct; 2 marks for 3–4 correct; 1 mark for 1–2 correct]

   d) Slope of graph = 5.1 ÷ 0.5 [1] = 10.2Ω [1]; The slope and the average of the ratio are essentially the same value [1]

e) Voltage = Current × Resistance, or V = IR **[1]**; Voltage in volts **[1]**; Current in amps **[1]**; Resistance in ohms **[1]**

f) The experiment can be repeated **[1]**; The experimental results can be reproduced **[1]**

2. a)

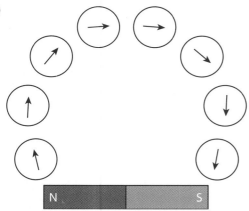

[2 marks: 1 for showing arrows on compass needles; 1 for showing the correct orientation of arrows; –1 for each error or omission]

b)

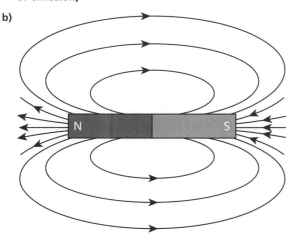

[3 marks: 1 for showing at least six field lines; 1 for including arrows; 1 for showing the correct orientation of arrows]

c) Sprinkling iron filings over a paper covering a bar magnet **[1]** and tapping gently **[1]** (or similar)

d) The closeness of the iron filings (indicate a stronger magnetic field) **[1]**

e) In a bar magnet the magnetic domains are aligned in the same direction **[1]**. When a magnet is split in two, each smaller magnet still retains its magnetic domains **[1]**. The number of domains in each smaller magnet is also reduced **[1]**; hence, the two smaller magnets have a smaller magnetic field strength **[1]** (or similar)

Illustrative image (optional) should look similar to this:

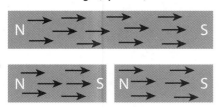

Science in Use

1. a)

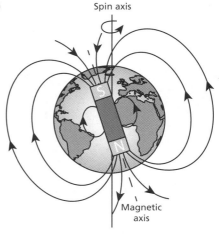

[3 marks: 1 for correct N and S poles; 1 for showing at least two field lines; 1 for showing arrows correctly orientated; Accept diagrams that show distorted field lines on one side]

b) south **[1]**; north **[1]**; south **[1]**, attract **[1]**

c) i) The solar wind **[1]**
   ii) It prevents deadly cosmic rays **[1]** and highly charged particles **[1]** from reaching the Earth

d) Mercury is smaller than the Earth (it has a smaller mass), so it has a weaker magnetic field **[1]**; A weak magnetic field tells us that Mercury may have a rotating **[1]** liquid metal core **[1]**

e) **Accept any three from:** Loudspeakers; Microphones; MRI scanners; Accelerators; Magnetic bearings; Magnetic disc players; Dynamos; Motors; Alternators **[3]**

a) Why is ice floating on water unusual when compared to most other substances?

....

....

.... [2]

b) What will happen to icebergs if the atmosphere continues to warm up?

.... [1]

c) Is this a **chemical** or **reversible** process?
Explain your answer.

....

.... [2]

d) What could happen to sea levels if the atmosphere continues to warm up?
Explain your answer.

....

....

.... [2]

2 When our body gets too hot we start to sweat. This helps to cool us down. Sweat is released by the body and as it evaporates from the skin we cool down.

a) What change of state occurs during evaporation?

.... [1]

b) What happens to the energy of the particles during evaporation?

.... [1]

c) What is the energy called that enables liquids to evaporate?

.... [1]

d) People often spray perfume, which is a liquid, but the pleasant aroma of perfume vapour (gas) can be detected without being right next to the wearer.

i) How does liquid perfume turn into a gas?

.... [1]

ii) Where does the energy come from to enable perfume to turn into a vapour?

.... [1]

iii) What name is given to the process by which perfume particles travel through the air so that they can be smelled at the other side of the room?

.... [1]

Total Marks .... / 13

	Vocabulary Builder	Maths Skills	Testing Understanding	Working Scientifically	Science in Use
Total Marks	/ 13	/ 13	/ 39	/ 10	/ 13

Chemistry — Explaining Chemical Changes

Vocabulary Builder

1 For each of the following statements, say whether it applies to an **acid**, an **alkali** or **salt**.

a) Has a pH of less than 7 .... [1]

b) Is neutral .... [1]

c) Turns universal indicator blue .... [1]

d) Is formed during a neutralisation reaction .... [1]

2 Underline the statement that does **not** describe an **acid**.

A: Sour tasting

B: Corrosive or irritant

C: Contains hydrogen

D: Has a pH of more than 7 [1]

3 Which of the following statements about an **alkali** are true (**T**) and which are false (**F**)? Write **T** or **F** in the spaces provided.

a) Soapy to touch .... [1]

b) Turns litmus red .... [1]

c) Contains hydroxide ions .... [1]

d) Corrosive .... [1]

4 Underline the statement that does **not** describe a **neutralisation** reaction.

A: Vinegar mixing with baking powder

B: Hydrochloric acid mixing with sodium hydroxide

C: Indigestion tablets in the stomach

D: Water mixing with sulfuric acid [1]

5 What do **catalysts** do?

....

.... [2]

6 Which one of the following is a **combustion reaction**?
Underline the correct option.

    A: An acid reacting with metal

    B: Petrol reacting with oxygen

    C: Water boiling

    D: Iron rusting   [1]

7 This question is about chemicals used in school laboratories.

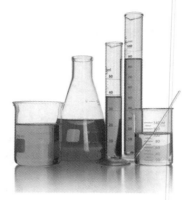

Complete the sentences below by filling in the missing words.

Common chemicals in the school laboratory include hydrochloric, nitric and sulfuric

_____. These chemicals all contain the element _____ and

have a pH less than _____. Laboratory alkalis include sodium

_____ and all alkalis have a pH of more than _____.

A base is the name for any chemical that will _____ with an acid, and an

alkali is a _____ base. When an acid and a base react together a

_____ is always formed.   [8]

8 One of the following word equations does **not** represent the reaction of an acid.
Underline which one.

    A: Acid + Metal $\longrightarrow$ Salt + Hydrogen

    B: Acid + Alkali $\longrightarrow$ Salt + Water

    C: Acid + Water $\longrightarrow$ Salt + Alkali

    D: Acid + Metal oxide $\longrightarrow$ Salt + Water   [1]

Total Marks _____ / 22

1 A student added 10g of marble chips (calcium carbonate) to a conical flask containing 50cm³ of hydrochloric acid on a balance, as shown in the diagram.

During the experiment, carbon dioxide gas is given off and enters the atmosphere.

The graph below shows the mass reading on the balance every minute for 20 minutes.

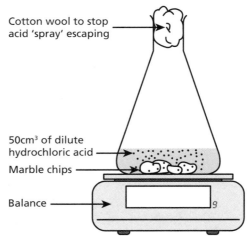

Cotton wool to stop acid 'spray' escaping

50cm³ of dilute hydrochloric acid

Marble chips

Balance

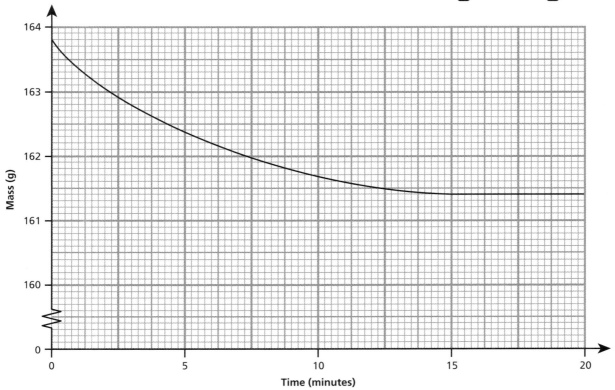

a) What is the starting mass in this experiment? .... [2]

b) What is the mass of the beaker, acid and marble chips at the end of the experiment? .... [2]

c) Calculate the mass of carbon dioxide given off in this experiment. Show your working.

.... [2]

d) After how many minutes did the acid finish reacting with the marble chips? Explain your answer.

....

.... [2]

2 The pie chart shows the main industrial uses of nitric acid (HNO_3).

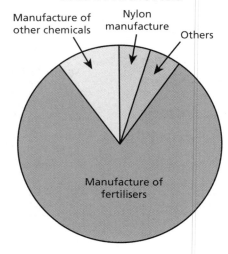

Uses of Nitric Acid

a) What is the major industrial use for nitric acid?

.... [1]

b) Approximately what percentage of nitric acid is used to manufacture other chemicals?

.... [1]

Total Marks .... / 10

Testing Understanding

1 Complete the word equations below by writing in the names of the missing chemicals.

a) Magnesium + Hydrochloric acid $\rightarrow$ Magnesium chloride + ... [1]

b) .... + Sulfuric acid $\rightarrow$ Zinc sulfate + Hydrogen [1]

c) Sodium hydroxide + Hydrochloric acid $\rightarrow$ Sodium chloride + ... [1]

2 Complete the diagram of the fire triangle by writing in the two missing requirements for combustion to occur.

A: ....   B: ....

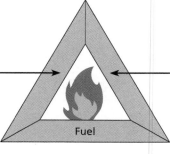

Fuel

[2]

3 **Indicators** are chemicals that can be used to determine whether substances are acids or alkalis.

The table shows the colours that indicators turn in different chemicals.

Indicator	Colour in Acid	Colour in Alkali
Litmus	Red	Blue
Phenolphthalein	Colourless	Pink
Bromothymol blue	Yellow	Blue

The table below shows the pH value of some common chemicals, as well as the colour that universal indicator will turn in different pH levels.

Stomach acid (pH 1) Lemon juice (pH 2) Cola (pH 3) Tomato juice (pH 4) Saliva (pH 6.5) Water (pH 7) Blood, tears (pH 7.5) Sea water (pH 8) Baking soda (pH 9) Household ammonia (pH 12)

pH	1	2	3	4	5	6	7	8	9	10	11	12	13	14
Colour	Red		Orange		Yellow		Green	Dark green		Blue		Indigo–violet		

Use the list of common chemicals to answer the following questions.

a) Which is the most acidic substance? .... [1]

b) Which is the most alkaline substance? .... [1]

c) Name the neutral substance. .... [1]

d) Name one substance that is more alkaline than sea water. .... [1]

4 Underline the statement that is **not** normally associated with a chemical change.

    A: Bubbles of gas being given off

    B: A change in temperature

    C: A colour change

    D: A reversible reaction [1]

5 Becky was investigating the reaction between hydrochloric acid and magnesium. She placed 10cm³ of hydrochloric acid into a boiling tube and added a piece of magnesium metal. When hydrochloric acid reacts with magnesium, a salt and hydrogen gas are formed.

a) Name the piece of apparatus Becky would have used to measure out the hydrochloric acid. .... [1]

b) Name a substance that Becky could have used to confirm that hydrochloric acid is an acid.

.... [1]

c) How would Becky have been able to confirm that hydrochloric acid was an acid?

.... [1]

d) Name the salt formed when magnesium reacts with hydrochloric acid.
   Circle the correct answer from the options below.

    magnesium oxide   **magnesium chloride**   **magnesium sulfate**   [1]

e) Describe the test and result that Becky would have used to show that hydrogen gas was produced in this experiment.

....

.... [2]

6 Complete the table below to show the name of the salt formed when different acids react with different metals. The first metal has been done for you.

Acid / Metal	Zinc	Iron	Calcium
Hydrochloric acid	Zinc chloride		
Sulfuric acid	Zinc sulfate		
Nitric acid	Zinc nitrate		

[6]

7 Complete the table below to show the name of the salt formed when different acids react with different alkalis. The first alkali has been done for you.

Acid / Alkali	Sodium Hydroxide	Calcium Hydroxide	Aluminium Hydroxide
Hydrochloric acid	Sodium chloride		
Sulfuric acid	Sodium sulfate		
Nitric acid	Sodium nitrate		

[6]

8 During a chemical reaction, particles in the reactants join in a different way to form the products of the reaction. The equation below represents the reaction between carbon and oxygen to form carbon dioxide.

$$C + O_2 \rightarrow CO_2$$

a) Carbon dioxide and oxygen exist as molecules. What does the word **molecule** mean?

_____ [1]

b) What does the 2 in O_2 represent?

_____ [1]

c) What are the reactants in the above reaction?

_____ and _____ [2]

d) What is the product in the above reaction?

_____ [1]

e) Write a word equation for the above reaction.

_____ [2]

Total Marks _____ / 35

Working Scientifically

1 Acid / alkali indicators can be made by extracting the pigment from certain fruit and vegetables such as red cabbage and blackberries.

The diagram below shows how red cabbage indicator can be made in the laboratory.

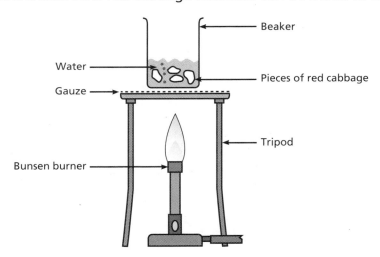

After a while, the water turns purple. Upon cooling, the mixture can be filtered and the liquid used as an indicator.

a) What is an **indicator**?

....

.... [1]

b) Why should the liquid be cooled before filtering?

.... [1]

c) Red cabbage indicator turns red in acidic conditions and green in alkaline conditions.

   i) Complete the table below to show the colour that red cabbage indicator will turn for different household chemicals.

Household Chemical	Colour in Litmus	Colour in Red Cabbage Indicator
Vinegar	Red	
Oven cleaner	Blue	

[2]

   ii) Explain how you can use red cabbage indicator to show that a solution is neutral.

.... [1]

Total Marks .... / 5

Science in Use

1 Read the passage about indigestion and then answer the questions that follow.

The acid in your stomach is **hydrochloric acid** and one of the symptoms of **indigestion** is having too much acid in your stomach. Indigestion tablets (known as **antacids**) contain chemicals that react with the acid and, therefore, they relieve the discomfort caused by indigestion.

A scientist was investigating the effectiveness of some indigestion tablets.
He placed 50cm^3 of acid into a beaker, followed by three drops of universal indicator.
He then added indigestion tablets until the indicator turned green.

a) What name is given to the reaction of acids with chemicals that remove their acidity?

.... [1]

b) Why is it important that scientists use a fixed amount of acid when comparing the effectiveness of indigestion tablets?
Explain your answer.

....

....

.... [2]

c) What colour will the universal indicator have turned when it was added to the beaker containing the acid?

.... [1]

d) Why did the scientist stop adding tablets when the universal indicator had turned green?

....

....

.... [2]

e) The results of this experiment are shown below.

Brand of Tablet	Number of Tablets Needed to Turn the Indicator Green
Tummycalm	7
Bellysoothe	4
Stomachnorm	6

Which brand of tablet is the most effective indigestion cure?
Explain your answer.

....

....

.... [2]

2 Water companies need to check the level of acidity in drinking water to ensure that when it reaches our homes it is not too acidic or alkaline. Scientists regularly carry out reactions called **titrations** to check the level of acidity in drinking water.

The diagram below shows how a titration can be carried out.

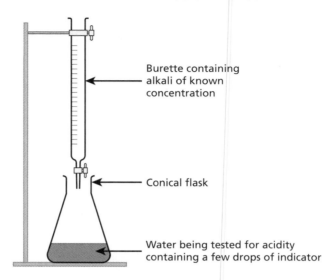

Burette containing alkali of known concentration

Conical flask

Water being tested for acidity containing a few drops of indicator

A known quantity of drinking water is placed into the conical flask and some indicator is added. An alkali is then placed in the **burette** and slowly added to the water until the indicator shows that any acid has been neutralised. The burette allows alkali to be added to the water one drop at a time. It also allows the volume of alkali added to the water to be accurately recorded.

a) Why is it important that the level of acidity in drinking water is regularly tested?

_____

_____ [2]

b) In what two ways is a burette an accurate piece of apparatus?

_____

_____ [2]

c) How will the scientist know when all of the acid has been neutralised?

_____ [1]

Total Marks _____ / 13

	Vocabulary Builder	Maths Skills	Testing Understanding	Working Scientifically	Science in Use
Total Marks	/ 22	/ 10	/ 35	/ 5	/ 13

1 Complete the following passage by filling in the gaps with key words from the box.

static	electrically	voltage	current
discharged	insulating	conducting	

Some _____ materials can become _____ charged when

they are rubbed against each other. Unless it is _____, the charge (called

_____ electricity) stays on the material. This type of charge can build up

to form a large electrical _____ .   [5]

2 Use the words from the box to complete the following passage about the build-up and transfer of static electricity.

negative	electrons	loses	positive	attractive	unlike

When a polythene rod is rubbed with a cloth, _____ move from the cloth

onto the rod leaving the rod with a _____ charge and the cloth with a

_____ charge. When an acetate rod is rubbed, the rod becomes

positively charged because it _____ electrons. Bringing the acetate and

polythene rods close together provides an _____ force as

_____ charges attract.   [6]

3 Complete the following sentences by circling the correct words in bold.

a) The near spherical Earth acts as if its mass is concentrated at the **surface / centre** of
the Earth.   [1]

b) Any mass on the Earth's surface is therefore **attracted to / repelled by** the centre of
the Earth by a non-contact force called the **electrostatic / potential / gravitational** field. [2]

c) This field exerts a force we call **momentum / weight** on any mass in this field.   [1]

d) The value of the field strength is approximately $\frac{1}{6}$**th / 10 / 1000** N/kg at the
Earth's surface.   [1]

4. Which of the following statements are true (**T**) and which are false (**F**)?
Write **T** or **F** in the spaces provided.

a) If an object is submerged in a liquid, the deeper it
goes the greater the buoyancy. .... [1]

b) If an object is submerged in a liquid, the liquid
provides a buoyancy force. .... [1]

c) When an object is lowered into a liquid, its mass decreases. .... [1]

5. a) Mountaineers often use a small piece of equipment to record
their altitude. What is the name given to such equipment? .... [1]

b) Such equipment also has built-in devices to measure the
atmospheric pressure. What is the name given to these devices? .... [1]

c) What is the name of the unit that is normally associated
with measuring atmospheric pressure? .... [1]

d) As mountaineers go higher what happens to the air pressure?

.... [1]

e) Explain briefly, in terms of pressure, how surface winds are formed.

....

.... [2]

6. Choose the correct terms from the box below to complete the sentences.
Some terms may be used more than once or not at all.

weight	electrostatic field	friction	gravitational field

a) As a rocket leaves the Earth's atmosphere, the pulling force, called the

......................................, becomes weaker and weaker. [1]

b) A rocket also has to overcome atmospheric .... caused by the
air resistance within the atmosphere itself. [1]

c) The strongest .... is produced by the Sun because of its
massive size and mass. [1]

d) The Moon's .... is about one-sixth of the value here on Earth. [1]

e) The influence of .... on a spaceship is far less when it finally
leaves the Solar System. [1]

Total Marks .... / 30

1 Give the formula for pressure (P) in terms of the applied force (F) and the area (A). Specify the units. You can write the formula either in words or in symbols.

....

....

.... [4]

2 A large container is pressing on a concrete floor with a downward force of 20 000N. If the area of the container that is in contact with the floor is 16m^2, calculate the pressure exerted by the container.

....

....

.... [3]

3 A person of mass 65kg is standing on the floor. The shoes they are wearing cover an area of 0.32m^2. Calculate the pressure exerted on the floor by the person. Give your answer to 2 s.f.

....

....

....

.... [3]

4 For this question, take standard atmospheric pressure to be 100 000N/m^2.

a) Write down the value of standard atmospheric pressure in terms of hectopascals.

....

.... [2]

b) What force is exerted on a box of cross-sectional area 1.4m^2 due to atmospheric pressure?

....

.... [2]

c) What is this force equivalent to in terms of mass? (Take g = 10N/kg.)

....

.... [2]

Total Marks .... / 16

1 a) Name two principle forces that act at a distance (**non-contact forces**).

....

.... [2]

b) Name a principle force that is a **contact force**.

.... [1]

2 Draw lines to match the correct pairs of statements.

Gaining electrons		Positively charged
Negatively charged		Losing electrons

[2]

3 a) The diagram shows an inflated balloon hanging freely next to a wall by static electricity. Mark on the diagram the charges (+) and (–) that are involved in keeping the balloon in this position.

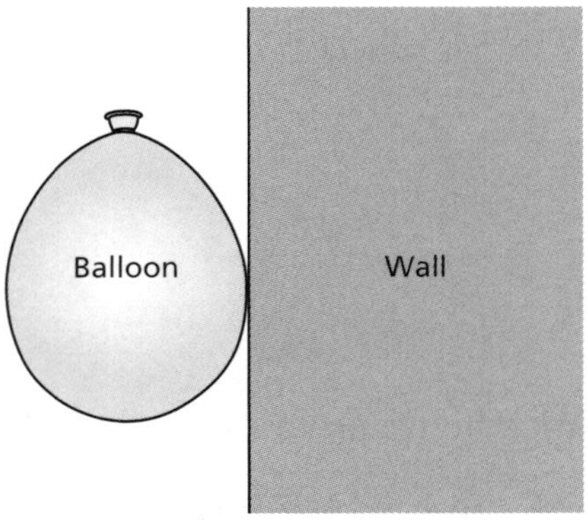

Balloon   Wall

[2]

b) A rubber balloon rubbed on a woollen jumper tends to 'stick' to the wall or ceiling. Explain in terms of static electricity why it stays there. Use the diagram above to help you.

....

....

....

....

.... [4]

4 What do we call the lines of force between two like (or unlike) charges that are not in contact?

_____ [1]

5 The gravitational field strength of the Earth is approximately 10N/kg. Choosing from the values in the box, state the approximate strength of the gravitational fields on:

a) the Moon ....

1.0N/kg	50.0N/kg	1.7N/kg
25.0N/kg	2.5N/kg	15.4N/kg

[1]

b) the planet Jupiter. .... [1]

6 What **key** physical quantities influence the strength of a gravitational field on a planet or moon?

_____ [2]

7 Which of the following statements are true (**T**) and which ones are false (**F**)?
Write **T** or **F** in the spaces provided.

a) Pressure in a liquid depends on the density. .... [1]

b) Pressure in a liquid acts in all directions. .... [1]

c) Pressure in a liquid only acts in the downward direction. .... [1]

d) Pressure in a liquid is transmitted in all directions. .... [1]

e) Pressure in a liquid decreases with depth. .... [1]

8 For each of the following statements, explain briefly what is happening in terms of pressure.

a) Walking over soft snow in snow shoes

_____

_____ [2]

b) Aeroplanes flying with pressurised cabins

_____

_____ [2]

c) Deep sea divers wearing strong and heavy diving suits

_____

_____ [2]

Total Marks .... / 27

1 A school experiment was undertaken to investigate the connection (if any) between water pressure and water depth using a plastic cylinder filled with water. Holes were made at 5cm intervals from the top of the cylinder. It was then filled with water.

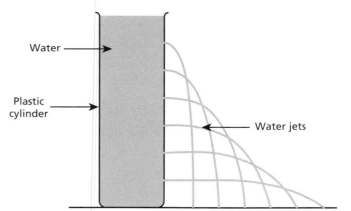

The horizontal distance the water jetted from each of the holes was recorded with a ruler.

The results of the experiment are shown in the table.

Depth of Water (cm) (i.e. Distance from the Water's Surface)	Horizontal Distance Travelled by the Water Jet (cm)
5	1.4
10	3.1
15	4.3
20	5.2
25	7.4
30	8.1

a) Using the data in the table, plot a graph of the depth of water against the recorded horizontal distance travelled by the water jet.
   Draw the best straight line through your plotted points.

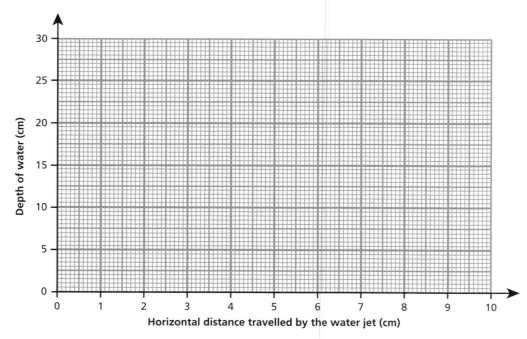

[4]

b) What causes the water jets to travel different horizontal distances from each of the holes?

_____ [1]

c) Is there a connection or relationship between the horizontal distance travelled by the water jet and the depth of water?
If so, state what it might be.

_____

_____

_____ [2]

d) At any point within the cylinder, the pressure is the same in all directions.
When using water, the pressure is given by the formula:

Pressure (P) = 1000 × Gravitational field strength (g) × Depth of water (h)

where P has units of pascals, g has units of N/kg and h is in metres.
If the depth of water is 30cm, use the above formula to calculate the water pressure at this position. (Take g = 10N/kg.)

_____

_____

_____

_____ [3]

e) The above formula includes the term **1000**, which represents a quantity that has a value of $1000kg/m^3$ for water.
What is the name of this quantity?

_____ [1]

f) i) Deep sea divers often search for wrecks in water 30m deep.
Using the above formula, determine the pressure on a diver who is 30m below sea level.
Give your answer in kPa.

_____

_____ [2]

ii) If the atmospheric pressure at sea level is 100kPa, what is the answer to part **i)** in terms of atmospheric pressure? (Give your answer as a whole number.)

_____ [1]

Total Marks _____ / 14

1 Read the article below about gravity and then answer the questions that follow.

Gravity exits everywhere; it keeps the Earth revolving around the Sun and the remaining planets and moons in their orbits. It keeps our solar system together, as well as other solar systems around other stars and in other galaxies.

The **force** due to gravity does, however, vary with **distance**. This is clearly seen on the graph, which shows how the **gravitational field strength (g)** varies with the distance above the Earth's surface.

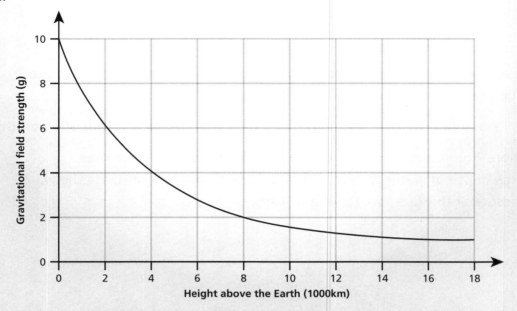

Although at far distances Earth's gravity is weak, the Sun's gravity is still strong enough to keep even the dwarf planet Pluto (which sits at the very edge of our solar system) in its orbit.

Without the force due to gravity, there would be no atmosphere on Earth and hence no air resistance. It is therefore a very important force that influences the motion of all objects on Earth and in space.

a) The strength of the Earth's gravitational field at the Earth's surface is 10
(as seen on the graph above).
What are the units of gravitational field strength?
Circle the correct answer.

  N   kg   N/kg   kg/N   m/s   [1]

b) Which of the following statements about the force due to gravity are true (**T**) and which ones are false (**F**)? Write **T** or **F** in the spaces provided.

   i) It is an attractive force. .... [1]

   ii) It acts between all objects. .... [1]

   iii) It only occurs for very large objects. .... [1]

   iv) It acts towards the centre of an object. .... [1]

   v) It decreases as the distance between objects increases. .... [1]

   vi) It is zero at some point in space. .... [1]

c) An unmanned vehicle is being launched towards the Moon.

   i) If the mass of the vehicle is 300kg on Earth, what will be its weight on the launch pad?

   ....

   ....

   .... [2]

   ii) Using the graph on page 56, work out the approximate weight of the vehicle when it is 18 000km above the Earth.

   ....

   ....

   .... [2]

   iii) What will be the weight of the vehicle when it lands on the Moon?
   (Take the Moon's gravitational field strength to be one-sixth of that on Earth.)

   ....

   ....

   .... [2]

   iv) Explain why there is a difference between your answers to parts **ii)** and **iii)**.

   ....

   ....

   ....

   .... [2]

d) When a pair of Apollo astronauts landed on the Moon, they carried out a gravity experiment using a hammer and a feather. The hammer and feather were both dropped at the same time.

Which of the following statements are true (**T**) and which are false (**F**)?
Write **T** or **F** in the spaces provided.

i) A hammer and a feather would land at the same time on Earth. .... [1]

ii) A hammer and a feather would land at the same time on the Moon. .... [1]

iii) A hammer would land before a feather on Earth. .... [1]

iv) A hammer would land before a feather on the Moon. .... [1]

v) A hammer would fall faster on Earth than on the Moon because 'g' is larger. .... [1]

e) The Voyager I spacecraft has effectively left our solar system and has entered interstellar space. Its motors are no longer effective.
Explain what major influence will govern its final flight path.

....

.... [2]

Total Marks .... / 22

	Vocabulary Builder	Maths Skills	Testing Understanding	Working Scientifically	Science in Use
Total Marks	.... / 30	.... / 16	.... / 27	.... / 14	.... / 22

1 a) Complete the following sentence by inserting the **common** key word.

Good electrical conductors have very low _____, whilst electrical

insulators have very high _____. [1]

b) Here is a list of some common materials and substances.

copper	plastic	iron	aluminium	rubber
	polystyrene	wood	paint	

Which of these materials and substances are good electrical conductors and which are good electrical insulators? Place them in the correct column in the table below.

Conductor	Insulator

[4]

2 Using the key words below, fill in the gaps to complete the passage about a magnetic material.

attract	iron	N-pole	domains	repel	S-pole

A magnetised piece of metal such as _____ contains small areas called

magnetic _____ that tend to point in the same direction. One end of the

magnet is collectivity termed the _____ and the other end the

_____. When two magnets are brought close together, like poles

_____ each other and unlike poles _____. [6]

3 Complete the following passage about the flow of electricity in a simple circuit by circling the correct words in bold.

The conventional electric **current / voltage** is the flow of **neutrons / protons / electrons** from **positive / negative** to **positive / negative**. It is measured in **volts / amperes / watts** using **a voltmeter / an ammeter / a power meter**. [6]

4 **a)** Describe the main difference between a **permanent magnet** and a **temporary magnet**.

....

....

....

....

....

.... [2]

b) Define what is meant by the term **electromagnet** and give two areas of application
where electromagnets are used.

....

....

....

....

....

.... [4]

5 Look at the series circuit below, which contains a battery and a light bulb.
Fill in the gaps to complete the following passage about this circuit.

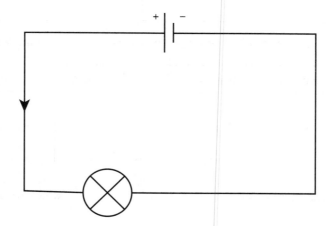

As a result of .... changes within a cell or battery, an electromotive force

is generated, which gives .... energy that pushes them around the circuit.

This force is measured in .... When electrons come into contact with a

lamp, they give up this energy to the thin wires in the lamp so that the wires become

.... and glow, giving off light. [4]

6 **a)** The diagram shows a cross-sectional view through the Earth.
   Label the diagram with the correct words.

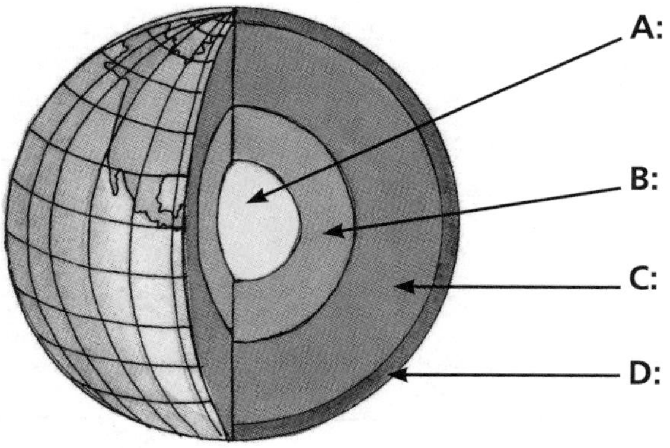

A: ....

B: ....

C: ....

D: ....

[4]

b) Describe briefly the **geodynamic theory**, with reference to the picture above.

....

....

....

....

....

[4]

Total Marks .... / 35

Maths Skills

1 A hairdryer draws a current of 5A.
   If the mains supply is taken to be 230V, calculate the resistance of the heater in the hairdryer.

....

....

....

[2]

2 An electric kettle has a metal heating element with a resistance of 23Ω.
   If the mains supply is 230 volts, calculate the current flowing through the heating element.

....

....

....

[3]

3 Two small light bulbs each with a resistance of 3Ω are connected in **series** to a 6 volt battery.

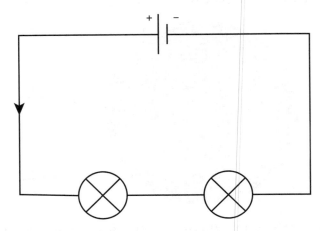

a) Calculate the current across the circuit.

....

....

.... [3]

b) Calculate the potential difference across each light bulb.

....

....

.... [2]

c) The two light bulbs are now replaced by one bulb with a resistance of 2 ohms and another bulb with a resistance of 4 ohms.
What is the potential difference across the 4 ohm bulb?

....

....

.... [2]

4 A 6 volt battery provides the potential difference for three light bulbs arranged in **parallel**. Each light bulb has a resistance value of 2 ohms.

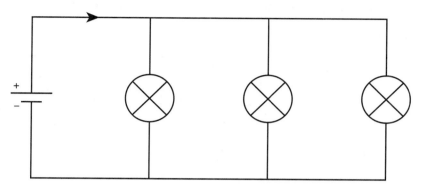

a) What is the potential difference across each of the bulbs?

.... [1]

b) Calculate the current across each of the bulbs.

....

....

....

.... [3]

Total Marks .... / 16

Testing Understanding

1 Look at the following diagram.

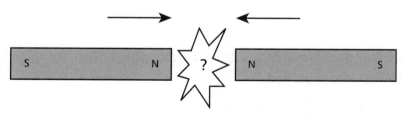

What effect will be produced if two similar poles of a bar magnet are brought closer together?

.... [1]

2 **a)** Look at circuits **A** to **F**, which show different arrangements of two light bulbs. Decide whether each circuit is a **series** circuit or a **parallel** circuit.

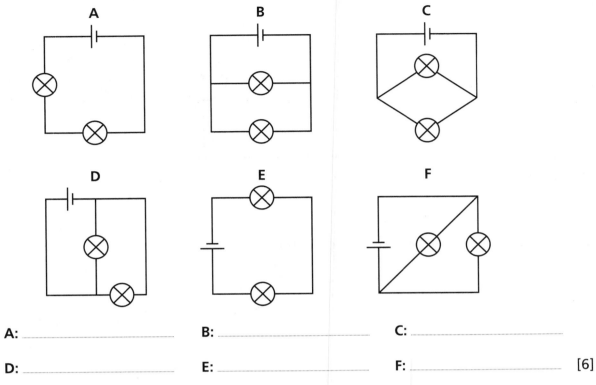

A: .... B: .... C: ....

D: .... E: .... F: .... [6]

b) If one of the bulbs 'blows' in these circuits, what would be the effect seen on the other bulb in the:

i) parallel circuit(s)

....[1]

ii) series circuit(s)?

....[1]

3 Which of the following statements are true (**T**) and which ones are false (**F**)? Write **T** or **F** in the spaces provided.

a) In a series circuit, the current decreases as it passes through each component. .... [1]

b) In a parallel circuit, the current adds where branches meet. .... [1]

c) In a parallel circuit, the current flows equally in all parts of the circuit. .... [1]

d) In a series circuit, the current is the same in all parts of the circuit. .... [1]

4 Look at the following diagram, which shows a wire carrying a current between two magnets.

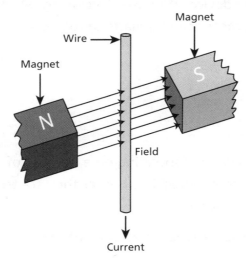

a) What effect will you observe when the wire carries **no** current?

.... [1]

b) What effect will you observe when a small current is passed through the wire?

.... [1]

c) Explain the effect observed in part **b)**.

....

....

....

.... [4]

d) What effect will you observe if a current is now passed through a loop of wire that is placed in the magnetic field?
   What is this effect called?

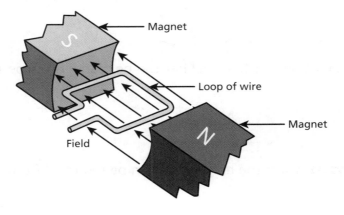

....

.... [2]

Testing Understanding (continued)

5 Electric motors are common devices that use electromagnets or permanent magnets.
Give three ways in which a motor can be made to rotate at an increased rate.

....

....

.... [3]

6 The resistance of a material varies depending on a number of factors.
Look at the statements below and decide whether they are true (**T**) or false (**F**).
Write **T** or **F** in the spaces provided.

a) Resistance decreases as the cross-sectional area of the material decreases. .... [1]

b) Resistance increases as the temperature of the material increases. .... [1]

c) Resistance increases as the length of the material increases. .... [1]

d) Resistance depends on the substance from which the material is made. .... [1]

7 The Earth has a magnetic field. How does this fact help you to navigate in the hills of the UK using only a map and a compass?

....

....

....

....

.... [3]

8 **a)** Give two methods by which a steel bar can be made into a permanent magnet.

....

.... [2]

b) Give two ways by which the magnetism can be removed from a permanent magnet.

....

.... [2]

Total Marks .... / 35

1 Students are using a standard circuit in an experiment to determine the resistance of a piece of material. The diagram shows a circuit containing a battery and the material of unknown resistance. An ammeter and a voltmeter also form part of the circuit and are used to measure the current and potential difference, respectively.

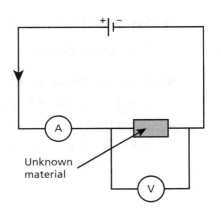

Unknown material

a) Complete the following sentence by circling the correct words in bold.

The ammeter is connected in **series / parallel**, whilst the voltmeter is connected in **series / parallel**. [2]

b) The students vary the current within the circuit using a rheostat (not shown) and they record the potential difference. The results of the experiment are shown in the table.

Ammeter Reading (A)	0.1	0.2	0.3	0.4	0.5	
Voltmeter Reading (V)	0.9	2.1	3.0	3.9	5.1	
Resistance (Ω)						Average:

Plot the results onto the graph below. Draw the best straight line through your plotted points.

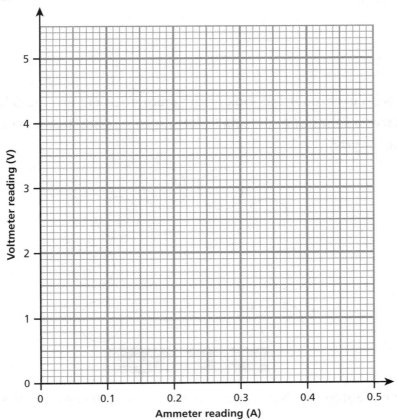

[4]

c) Complete the table on page 67 by calculating the ratio: $\dfrac{\text{voltmeter reading}}{\text{ammeter reading}}$

Give your values to 1 d.p., and find the average of this ratio. [3]

d) Calculate the slope of the straight line and compare this with the average value of the ratio you obtained above.

Given that there are uncertainties associated with the values in this experiment, comment briefly on how these two results compare.

....

....

.... [3]

e) Write the relationship in words or symbols that connects the resistance, the potential difference and the current in this circuit.

Give the name of the units associated with each part.

....

....

.... [4]

f) What two scientific methods can be used to test the reliability of this set of experimental data.

....

.... [2]

2 A class conducts an experiment to look at the magnetic field around a bar magnet. The teacher places a series of eight small plotting compasses around the bar magnet, as shown in the diagram.

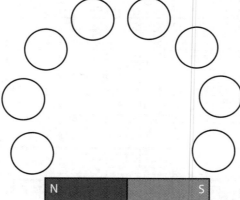

a) Complete the diagram by showing the direction of each compass needle. [2]

b) On the following diagram, draw the magnetic field shape around the bar magnet. You should show enough lines between the poles to clearly illustrate the effect.

| N | | S |

[3]

c) Describe a simple experiment (other than using plotting compasses) in which the magnetic field shape can be dramatically demonstrated and clearly seen.

[2]

d) How is the strength of the magnetic field shown in the experiment you described in part **c)**?

[1]

e) When a bar magnet is sliced in half, two smaller bar magnets are obtained possessing their own north and south poles. Briefly explain in terms of magnetic domains how this is possible. You may use diagrams to illustrate the effect.

[4]

Total Marks .... / 30

1 Read the passage about the Earth's magnetic field and then answer the questions that follow.

> The near spherical Earth spins on its axis once every 24 hours. The Earth is composed of a **solid inner core** and **liquid outer core**. It is the motion of this molten iron outer core due to the Earth's spin that generates a **magnetic field**, which can be visualised like a bar magnet at the centre of the Earth. Unlike a bar magnet, the motion of the molten iron varies, which means that the position of the north and south magnetic poles also varies. The magnetic north pole does wander in time, but it is slow enough (tens of thousands of years) that an ordinary compass remains a useful tool for navigation.
>
> There is now significant evidence to indicate that the north and south magnetic poles have changed places with each other at random intervals since the Earth was formed over 4 billion years ago. The average period between reversals is several hundred thousand years. This evidence has been obtained from the magnetic pole directions permanently recorded in rocks that were once molten, but have since cooled.
>
> The Earth's magnetic field extends several tens of thousands of kilometres into space, well above the ionosphere. This magnetic field that surrounds the Earth, called the **magnetosphere**, protects the Earth from harmful cosmic rays, the solar wind and strong ultraviolet radiation from the Sun that would eventually deplete the ozone layer and cause us harm. The shape of the magnetosphere is influenced largely by the solar wind.

a) On the following diagram of the Earth, the outline of a bar magnet is shown. Mark on the diagram the north and south poles of the bar magnet and show several lines of the magnetic field that surrounds the Earth (assume the Sun is to the right of the Earth). Indicate the direction of the field using arrows.

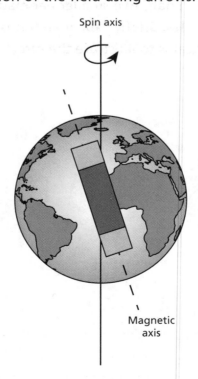

[3]

b) Fill in the gaps to complete the following passage, which describes the behaviour of a compass needle.

The north pole of a magnetic compass needle is also referred to as the Earth's magnetic

.... pole. It points in approximately the same direction as the Earth's

geographic .... pole because this is aligned roughly with the Earth's

magnetic .... pole, since opposite magnetic poles .... [4]

c) i) What influences the shape of the magnetosphere?

.... [1]

ii) What does it prevent?

....

.... [2]

d) The planet Mercury also possesses a magnetic field.
Suggest why Mercury's magnetic field is not as strong as the Earth's field.
What might this tell us about the structure of Mercury?

....

....

.... [3]

e) Magnets have found application in a number of areas.
Give three examples where magnets have been used to the benefit of humankind.

....

....

.... [3]

Total Marks .... / 16

	Vocabulary Builder	Maths Skills	Testing Understanding	Working Scientifically	Science in Use
Total Marks	.... / 35	.... / 16	.... / 35	.... / 30	.... / 16

The Periodic Table

Key

relative atomic mass
atomic symbol
name
atomic (proton) number

1	1 H hydrogen 1

1	2											3	4	5	6	7	0
																	4 **He** helium 2
7 **Li** lithium 3	9 **Be** beryllium 4											11 **B** boron 5	12 **C** carbon 6	14 **N** nitrogen 7	16 **O** oxygen 8	19 **F** fluorine 9	20 **Ne** neon 10
23 **Na** sodium 11	24 **Mg** magnesium 12											27 **Al** aluminium 13	28 **Si** silicon 14	31 **P** phosphorus 15	32 **S** sulfur 16	35.5 **Cl** chlorine 17	40 **Ar** argon 18
39 **K** potassium 19	40 **Ca** calcium 20	45 **Sc** scandium 21	48 **Ti** titanium 22	51 **V** vanadium 23	52 **Cr** chromium 24	55 **Mn** manganese 25	56 **Fe** iron 26	59 **Co** cobalt 27	59 **Ni** nickel 28	63.5 **Cu** copper 29	65 **Zn** zinc 30	70 **Ga** gallium 31	73 **Ge** germanium 32	75 **As** arsenic 33	79 **Se** selenium 34	80 **Br** bromine 35	84 **Kr** krypton 36
85 **Rb** rubidium 37	88 **Sr** strontium 38	89 **Y** yttrium 39	91 **Zr** zirconium 40	93 **Nb** niobium 41	96 **Mo** molybdenum 42	[98] **Tc** technetium 43	101 **Ru** ruthenium 44	103 **Rh** rhodium 45	106 **Pd** palladium 46	108 **Ag** silver 47	112 **Cd** cadmium 48	115 **In** indium 49	119 **Sn** tin 50	122 **Sb** antimony 51	128 **Te** tellurium 52	127 **I** iodine 53	131 **Xe** xenon 54
133 **Cs** caesium 55	137 **Ba** barium 56	139 **La*** lanthanum 57	178 **Hf** hafnium 72	181 **Ta** tantalum 73	184 **W** tungsten 74	186 **Re** rhenium 75	190 **Os** osmium 76	192 **Ir** iridium 77	195 **Pt** platinum 78	197 **Au** gold 79	201 **Hg** mercury 80	204 **Tl** thallium 81	207 **Pb** lead 82	209 **Bi** bismuth 83	[209] **Po** polonium 84	[210] **At** astatine 85	[222] **Rn** radon 86
[223] **Fr** francium 87	[226] **Ra** radium 88	[227] **Ac*** actinium 89	[261] **Rf** rutherfordium 104	[262] **Db** dubnium 105	[266] **Sg** seaborgium 106	[264] **Bh** bohrium 107	[277] **Hs** hassium 108	[268] **Mt** meitnerium 109	[271] **Ds** darmstadtium 110	[272] **Rg** roentgenium 111							

Elements with atomic numbers 112–116 have been reported but not fully authenticated

*The Lanthanoids (atomic numbers 58–71) and the Actinoids (atomic numbers 90–103) have been omitted.

Cu and **Cl** have not been rounded to the nearest whole number.